The Digital Highway
Revolutionizing Road Logistics with Technology

Orange Books Publication

1st Floor, Rajhans Arcade, Mall Road, Kohka, Bhilai, Chhattisgarh 490020

Website: **www.orangebooks.in**

First Edition, 2024

ISBN: 978-93-5621-808-6

THE DIGITAL HIGHWAY

REVOLUTIONIZING ROAD LOGISTICS WITH TECHNOLOGY

TRANSFORMING LOGISTICS FOR TOMORROW'S ROADS

DR. BALVINDER SINGH BANGA

OrangeBooks Publication

www.orangebooks.in

Dedication

This book is a reflection of the many influences that have helped shape it.

To my father, S. Kulwant Singh Banga who worked tirelessly in the railways, serving our nation and inspiring my own journey into logistics. His experiences and advice have guided me and given me a strong foundation for understanding the logistics industry.

To my beloved wife, Ritu Singh, whose unwavering support has been my greatest strength. Her love and devotion have been the steady anchor in our family, providing me with the peace and stability needed to pursue my passions. Through every challenge and triumph, she has stood by my side, creating a home filled with warmth, understanding, and endless encouragement. Her belief in me has not only made this journey possible but has also enriched our lives together in ways words cannot fully express. Without her, this book and everything it represents would not have been possible.

To my brother, Harry Banga, Chief Information Officer at CJ Darl, a leader in the 3PL sector. His knowledge of technology and logistics has broadened my view and sparked important discussions that have enriched this book.

Special thanks to my children, Kohinoor Singh Banga and Mansidak Singh Banga, whose love and belief in me inspire me every day. They give me purpose and push me to contribute meaningfully to the logistics field.

To the drivers and everyone in the logistics community, your hard work is the backbone of our economy. Your experiences have brought real-life insights and authenticity to this book.

I also want to express my deep gratitude to the government for launching the PM Gati Shakti scheme, the National Logistics Policy, and the Unified Logistics Interface Platform (ULIP). These initiatives are transforming India's infrastructure, driving innovation, efficiency, and sustainable growth. They show how important the partnership between industry and government is in shaping the future of logistics in our country.

This book is a tribute to everyone who has contributed to road logistics, leaving a lasting impact that will continue to shape our world.

Introduction

The transportation industry is experiencing a seismic transformation in an era when technology is permeating every aspect of our lives. The integration of digital advances and logistical frameworks is not just an incremental improvement but a complete revolution that will transform how goods are moved and managed. The road logistics sector is in the midst of unprecedented change, with autonomous vehicles traversing complex urban landscapes and sophisticated algorithms optimizing delivery routes.

This book delves into the myriad ways in which cutting-edge technology is redefining road logistics. It examines the convergence of artificial intelligence, machine learning, Internet of Things (IoT), and blockchain, and how these technologies are being harnessed to create smarter, more efficient, and more reliable transportation systems. The narrative unfolds through a blend of theoretical insights and practical applications, offering a comprehensive overview of the current landscape and a glimpse into the future.

As the global economy becomes increasingly interconnected, the demand for faster, more efficient, and cost-effective logistics solutions has never been greater. Traditional methods, while reliable, cannot keep pace with the accelerating speed of commerce and consumer expectations. Here, technology steps in as the game-changer, providing innovative solutions that address these burgeoning demands.

The book also explores the socio-economic implications of this digital transformation. While the benefits are many, including reduced operational costs, enhanced safety, and minimized environmental impact, the shift also brings challenges. Issues such as cybersecurity, data privacy, and the potential displacement of human labor are critically examined.

Through expert analysis, case studies, and interviews with industry leaders, this book provides readers with a holistic understanding of how technology is not just enhancing but revolutionizing road logistics. Whether you are a seasoned industry professional, a technology enthusiast, or a curious reader, this exploration offers valuable insights into the future of transportation and the digital innovations driving it forward.

Contents

Chapter 1

Introduction to the Digital Highway

Overview of Road Logistics- A Journey from Bullocks to Blockchain

Dust swirled on the ancient trade routes of India, where commerce thrived under the relentless march of time. The arteries of early economic development pulsed with the steady flow of goods, weaving through the subcontinent's vast expanse. These paths bore witness to the relentless pursuit of efficiency in transportation, an endeavor as old as civilization itself.

Through the haze of heat that danced upon the rocky terrain, bullock carts trundled along, laden with spices and textiles. Their wooden wheels creaked, a chorus in the symphony of trade that resonated across generations. Men shouted commands to their beasts of burden, voices rough with urgency.

"Arre! Hato!" one called out, halting his oxen to avoid a rut in the path. He wiped sweat from his brow with the back of his hand, his skin baked to the color of the earth he treads upon.

Another trader hoisted sacks over his shoulders, muscles straining under the weight. His feet sunk into the soft ground beside the road, each step a testament to the labor-intensive reality of his livelihood. The scent of musk and dust clung to the air, mingling with the distant sweetness of jasmine from a nearby grove.

"Careful there, the load must not shift," an elder cautioned, his voice carrying the authority of experience. He supervised the loading process, ensuring each item found its rightful place.

The scene was a mosaic of human toil and animal stamina. Each cart a vessel of hope, every journey a gamble against the elements and the caprices of the road. The traders knew the value of their cargo, understood the lifeline it represented for villages and cities alike.

"Without these routes, where would we be?" a young porter mused aloud, pausing to catch his breath.

"Without these roads, we are, but isolated islands in a vast sea," replied his companion, securing a rope around a stack of clay pots.

In this dance of supply and demand, the early days of road logistics. Progress marched on, one determined step at a time, setting the stage for a future where technology and tradition would collide, reshaping the contours of an age-old industry.

The story of road transport technology is a fascinating one, marked by continuous innovation.

The bullock cart and carriages dominated goods transportation for millennia. Standardization of parts and the internal combustion engine in the late 19th century laid the groundwork for future advancements.

And now road logistics becomes an integral component of the broader logistics industry, encompasses the planning, execution, and management of the transportation of goods via road networks.

This subchapter provides an in-depth examination of the key aspects of road logistics, highlighting its significance, operational dynamics, and the impact of digital technologies on its evolution.

Road logistics primarily involves the movement of goods using various types of vehicles, such as bullock cart, trucks, vans, and motorcycles. This mode of transport is characterized by its flexibility, accessibility, and cost-effectiveness, making it a preferred option for short to medium-distance deliveries. The infrastructure supporting road logistics includes an extensive network of highways, arterial roads, and urban streets, which facilitates the efficient movement of goods from production sites to consumption points.

The operational framework of road logistics is complex, involving multiple stakeholders, including shippers, carriers, logistics service providers, and regulatory authorities. Shippers, typically manufacturers or wholesalers, are responsible for the production and dispatch of goods. Carriers, such as trucking companies, handle the physical transportation of these goods. Logistics service providers offer a range of services, from warehousing and inventory management to route planning and freight forwarding. Regulatory authorities establish and enforce guidelines to ensure safety, efficiency, and environmental compliance within the industry.

Technological advancements have significantly transformed road logistics, enhancing its efficiency and reliability. Digital technology integration, such as telematics, GPS tracking, and IoT devices, has revolutionized fleet management and real-time monitoring. Using telematics systems, carriers can optimize their operations and reduce costs by obtaining comprehensive data on vehicle performance, driver behavior, and fuel consumption. GPS tracking ensures precise location information, facilitating accurate delivery estimates and improved route planning.

The advent of digital platforms and software solutions has further streamlined various aspects of road logistics. Transportation Management Systems (TMS) offer end-to-end visibility and control over the logistics process, from order placement to final delivery. These systems enable efficient load planning, route optimization, and carrier selection, thereby enhancing operational efficiency and customer satisfaction. Additionally, digital freight marketplaces connect shippers with carriers, enhancing market transparency and competitiveness.

The implementation of autonomous vehicles and advanced driver-assistance systems (ADAS) represents significant technological leap in road logistics. Autonomous trucks, equipped with sophisticated sensors and artificial intelligence promise to revolutionize the industry by reducing labor costs, minimizing human error, and enhancing safety. ADAS technologies, such as adaptive cruise control, lane-keeping assistance, and collision avoidance systems, improve driver safety and reduce accident rates.

Sustainability is a growing concern within road logistics, driven by environmental regulations and societal expectations. The industry is increasingly adopting eco-friendly practices, such as the use of alternative fuels, electric vehicles, and energy-efficient driving techniques. The shift towards greener logistics is supported by advancements in battery technology, renewable energy sources, and government incentives aimed at reducing carbon emissions.

Digitalization has also facilitated the development of smart logistics networks, characterized by interconnected systems and data-driven decision-making. Big data analytics and artificial intelligence enable predictive maintenance, demand forecasting, and dynamic route optimization, enhancing the resilience and responsiveness of road logistics operations. Blockchain technology offers potential benefits in terms of transparency, security, and traceability, particularly in complex supply chains involving multiple parties.

In conclusion, road logistics is a dynamic and evolving sector, significantly influenced by technological innovations and sustainability considerations. Digital technologies are reshaping traditional logistics models, driving efficiency, transparency, and environmental responsibility. As the industry continues to evolve, the integration of advanced digital solutions will be paramount in addressing emerging challenges and capitalizing on new opportunities.

Technological Evolution in Logistics

The logistics sector has undergone a significant transformation due to advancements in technology, fundamentally altering the landscape of supply chain management. Historically, logistics relied heavily on manual processes and human intervention, often leading to inefficiencies and errors. However, the digital revolution has introduced a plethora of tools and systems that streamline operations, enhance accuracy, and optimize resource utilization.

One of the most impactful innovations is the integration of Internet of Things (IoT) devices within the logistics framework. IoT devices, such as sensors and GPS trackers, provide real-time data on the location, condition, and status of goods in transit. This constant flow of information allows for better monitoring and management of shipments, reducing the risk of loss or damage. Furthermore, IoT-enabled predictive maintenance can foresee potential equipment failures, thus minimizing downtime and ensuring the continuity of operations.

Another critical development is the adoption of advanced analytics and big data. These technologies enable logistics companies to research vast amounts of information generated from various sources, including IoT devices, customer interactions, and market trends. By leveraging machine learning algorithms and **AI** (AI), firms can predict demand patterns, optimize route planning, and improve inventory management. This data-driven approach not only enhances efficiency but also provides a competitive edge in a rapidly evolving market.

Warehouse automation represents a significant leap forward in logistics technology. Automated storage and retrieval systems (AS/RS), along with autonomous mobile robots (AMRs), have revolutionized the way goods are handled and stored. These systems increase throughput, reduce labor costs, and minimize human error. Additionally, the implementation of robotics in warehouses facilitates faster order fulfillment and improves overall operational efficiency.

Blockchain technology has emerged as a game-changer in ensuring transparency and security within the supply chain. By creating a decentralized and immutable ledger, blockchain enhances traceability and accountability of goods throughout their journey. This technology mitigates risks associated with fraud and counterfeiting, providing stakeholders with a verifiable record of each transaction. Moreover, smart contracts—self-executing contracts with the terms directly written into code—automate and enforce contractual agreements, further streamlining logistics processes.

The role of cloud computing cannot be overlooked in the technological evolution of logistics. Cloud-based platforms offer scalable and flexible solutions for managing logistics operations. These platforms facilitate seamless collaboration and information sharing among different stakeholders, including suppliers, manufacturers, and customers. The ability to access real-time data from anywhere in the world enhances decision-making and agility, enabling companies to respond swiftly to changing market conditions.

Drones and autonomous vehicles are also making significant inroads into the logistics industry. Drones are being used for last-mile delivery, particularly in remote or hard-to-reach areas. They offer a cost-effective and efficient solution for delivering small packages quickly. Autonomous vehicles, on the other hand, hold the potential to revolutionize long-haul transportation. By reducing the reliance on human drivers, these vehicles can operate continuously, increasing efficiency and reducing delivery times.

The convergence of these technologies is driving the logistics sector towards a more interconnected and intelligent ecosystem. Companies that leverage these advancements are better positioned to meet the demands of modern consumers, who expect faster, more reliable, and transparent services. As technology continues to evolve, the logistics industry will undoubtedly witness further innovations that will shape the future of supply chain management.

Purpose and Scope of the Book

This book aims to provide a comprehensive exploration of the digital landscape, examining the transformative impact of digital technology on various facets of society, industry, and individual lives. It seeks to present an in-depth analysis of the historical development, current state, and future potential of digital innovations. The text aims to bridge the gap between technical intricacies and accessible knowledge, making it a valuable resource for both experts and laypersons interested in understanding the profound changes brought about by digital technology.

The scope of the book encompasses a wide array of topics, structured to offer a holistic view of the digital ecosystem. It begins with a historical overview, tracing the origins and evolution of digital technology from its nascent stages to its current pervasive presence. This section aims to provide context and background, highlighting key milestones and technological breakthroughs that have shaped the digital age.

Subsequently, the book delves into the core components of digital technology, including hardware, software, networks, and data. Each of these elements is examined in detail to elucidate their functions, interconnections, and the role they play in the broader digital infrastructure. Special emphasis is placed on understanding how these components work together to enable the seamless operation of digital systems.

The book also addresses the societal implications of digital technology, exploring its impact on various sectors such as healthcare, education, finance, and governance. Through case studies and real-world examples, it illustrates how digital innovations have revolutionized these fields, leading to increased efficiency, new opportunities, and unforeseen challenges. Ethical considerations and the digital divide are also discussed, underscoring the need for inclusive and responsible digital development.

Another critical area of focus is the economic dimension of the digital revolution. The book examines the ways in which digital technology has reshaped business models, market structures, and labor dynamics. Topics such as e-commerce, digital marketing, and the gig economy are analyzed to provide insights into the new economic paradigms emerging in the digital era. Additionally, the text explores the concept of digital transformation and its implications for organizational strategy and competitiveness.

Cybersecurity and privacy are addressed as pivotal concerns in the digital age. The book investigates the threats posed by cyber-attacks, data breaches, and digital surveillance, and discusses the measures needed to safeguard information and protect user privacy. It also considers the regulatory and policy frameworks that govern digital activities, highlighting the role of legislation in maintaining a secure and fair digital environment.

Emerging technologies such as artificial intelligence, blockchain, and the Internet of Things are explored in dedicated chapters. These sections aim to demystify advanced digital innovations and assess their potential to drive future developments. The book evaluates the opportunities and risks associated with these technologies, providing a balanced perspective on their transformative capabilities.

The concluding sections of the book project future trends and scenarios in the digital domain. By analyzing current trajectories and potential disruptions, it offers informed predictions about the direction of digital technology and its long-term implications for society. This forward-looking approach is intended to equip readers with the knowledge and foresight needed to navigate the evolving digital landscape.

Overall, the book aspires to be an authoritative guide on digital technology, offering a thorough and nuanced understanding of its multifaceted impact. It aims to inform, educate, and inspire readers, fostering a deeper appreciation of the digital age and its significance.

Methodology and Approach

During my doctorate, I conducted the research presented in this book, which employs a multifaceted methodology designed to address the complexities and nuances of the digital transformation landscape. This chapter delineates the methodological framework and the specific approaches employed to gather, analyze, and interpret data pertinent to understanding the various dimensions of the digital highway.

The study adopts a mixed-methods approach, integrating both qualitative and quantitative techniques to provide a comprehensive analysis. Quantitative data was primarily collected through structured surveys distributed to a diverse range of stakeholders, including industry experts, business leaders, and technology users. The surveys were meticulously designed to capture a wide array of variables, such as the adoption rates of digital technologies, the perceived benefits and challenges, and the economic impacts of digital transformation. Statistical tools and software, such as SPSS and R, were utilized to analyze the survey data, ensuring robust and reliable results.

In parallel, qualitative data was gathered through in-depth interviews and focus groups. These methods were chosen to gain deeper insights into the subjective experiences and perspectives of individuals directly involved in or affected by digital transformation initiatives. The interviews were semi-structured, allowing for flexibility in exploring emerging themes while maintaining a consistent line of inquiry. Focus groups, consisting of 5-8 participants, facilitated dynamic discussions and the generation of rich, contextual data. Transcriptions of these sessions were subjected to thematic analysis using NVivo software, which enabled the identification and categorization of recurring patterns and themes.

To complement the primary data collection, a comprehensive review of existing literature was conducted. This review encompassed academic journals, industry reports, white papers, and case studies, provided a theoretical foundation and contextual background for the study. The literature review also helped to identify gaps in the current knowledge base and informed the development of research questions and hypotheses.

In addition to primary and secondary data collection, the study incorporated case study methodology to examine specific instances of digital transformation in various sectors. The case studies selected were based on criteria such as industry relevance, geographic diversity, and the scale of digital initiatives. Detailed case analyses provided illustrative examples of best practices, challenges, and outcomes, offering practical insights and lessons learned.

The methodological rigor of the study was further enhanced by employing triangulation techniques. By cross-verifying data from multiple sources and methods, the research ensured greater validity and reliability of the findings. Triangulation also facilitated a more nuanced understanding of the digital transformation phenomena, capturing both macro-level trends and micro-level intricacies.

Ethical considerations were meticulously addressed throughout the research process. Informed consent was obtained from all the participants, and confidentiality was rigorously maintained. The study adhered to ethical guidelines stipulated by relevant institutional review boards, ensuring that the research was conducted with integrity and respect for participants' rights.

Data analysis involved both descriptive and inferential statistics for quantitative data, coding and thematic analysis for qualitative data. Descriptive statistics provided an overview of the key trends and patterns, while inferential statistics, such as regression analysis and hypothesis testing, allowed for the examination of relationships and causality. The qualitative data analysis focused on identifying themes, patterns, and narratives that elucidated the human and organizational dimensions of digital transformation.

The methodological approach outlined in this chapter underscores the study's commitment to rigor, comprehensiveness, and ethical integrity. By integrating multiple methods and sources of data, the research offers a robust and nuanced understanding of the digital highway, contributing valuable insights to both academic scholarship and practical applications in the field.

Chapter 2

Internet of Things (IoT) in Road Logistics

Basics of IoT

The Internet of Things (IoT) represents a transformative paradigm shift, connecting physical objects to the digital world through interconnected networks. This integration facilitates the exchange of data and commands, enabling enhanced automation, efficiency, and intelligence in various applications. The core concept of IoT revolves around embedding sensors, actuators, and communication modules into everyday objects, ranging from household appliances to industrial machinery, thereby creating a seamless web of interconnected devices.

At the heart of IoT is the notion of connectivity. Devices within an IoT ecosystem communicate through a variety of network protocols, including Wi-Fi, Bluetooth, Zigbee, and cellular networks. These protocols ensure that data can be transmitted reliably and securely across different environments and conditions. The choice of protocol often depends on the specific requirements of the application, such as range, power consumption, and data transfer rates.

Sensors play a crucial role in IoT by collecting data from the physical environment. These sensors can measure a wide array of parameters, including temperature, humidity, light, motion, and pressure. The collected data is then processed and analyzed to derive meaningful insights. For instance, in a smart home setup, temperature sensors can monitor ambient conditions and adjust the thermostat accordingly to maintain optimal comfort levels.

Actuators complement sensors by performing actions based on the data received. In the context of IoT, actuators can control various mechanical or electronic devices. For example, an actuator in a smart irrigation system can open or close valves to regulate water flow, based on soil moisture readings. This dynamic interaction between sensors and actuators forms the foundation of IoT's capability to automate and optimize processes.

Data processing and analysis are critical components of IoT. The vast amounts of data generated by IoT devices require robust processing capabilities to extract valuable information. Edge computing and cloud computing are two primary approaches used for this purpose. Edge computing involves processing data locally, near the source of data generation, which reduces latency and bandwidth usage. In contrast, cloud computing leverages centralized data centers to perform complex computations and store large datasets, providing scalability and advanced analytics.

Security and privacy are paramount concerns in IoT due to the extensive connectivity and data exchange involved. Ensuring the integrity, confidentiality, and authenticity of data is essential to protect against cyber threats. Various security measures, such as encryption, authentication, and access control, are implemented to safeguard IoT systems. Additionally, privacy considerations are addressed by implementing data anonymization techniques and adhering to regulatory frameworks.

Interoperability is another critical aspect of IoT, given the diverse range of devices and manufacturers involved. Standardization efforts, such as the development of communication protocols and data formats, aim to ensure seamless interaction between different devices and systems. Organizations like the Internet Engineering Task Force (IETF) and the Institute of Electrical and Electronics Engineers (IEEE) play a significant role in establishing these standards.

IoT has a wide array of applications across various sectors, including healthcare, agriculture, transportation, and manufacturing. In healthcare, IoT devices enable remote patient monitoring, facilitating timely medical interventions and personalized care. In agriculture, IoT systems optimize resource usage through precision farming techniques, enhancing crop yield and sustainability. The transportation sector benefits from IoT through smart traffic management and vehicle-to-vehicle communication, improving safety and efficiency. In manufacturing, IoT-driven automation and predictive maintenance enhance productivity and reduce downtime.

The integration of IoT into everyday life is reshaping industries and driving innovation. As technology continues to evolve, the potential of IoT to create smarter, more connected environments becomes increasingly apparent, laying the groundwork for a future, where digital and physical realms converge seamlessly.

IoT Applications in Transportation

The integration of the Internet of Things (IoT) within the transportation sector has initiated a transformative shift, fundamentally altering, how systems operate and interact. IoT applications in transportation span across various domains, including traffic management, vehicle-to-vehicle (V2V) and vehicle-to-infrastructure (V2I) communication, logistics, and public transportation, thereby enhancing efficiency, safety, and user experience.

Traffic management systems have become significantly more sophisticated with IoT. Sensors and connected devices collect real-time data on traffic flow, vehicle speed, and road conditions, enabling dynamic traffic signal control and congestion management. This data-driven approach facilitates adaptive traffic light systems that can respond to actual traffic conditions rather than pre-set schedules, reducing congestion and improving travel times. Furthermore, IoT-based traffic management systems can provide real-time updates to drivers through navigation apps, allowing for optimized route planning and avoidance of traffic jams.

V2V and V2I communication are pivotal IoT applications that enhance road safety and efficiency. Vehicles equipped with IoT devices can communicate with each other and with infrastructure elements such as traffic lights, road signs, and toll booths. This communication enables vehicles to share information about their speed, location, and direction, helping to prevent collisions and improve traffic flow. For example, V2V communication allows for coordinated braking in case of sudden stops, while V2I communication can provide drivers with warnings about upcoming traffic conditions or hazards.

In logistics, IoT applications are revolutionizing supply chain management and freight transportation. IoT-enabled tracking systems provide real-time visibility of goods in transit, allowing for precise monitoring of location, temperature, and condition of the cargo. This level of transparency enhances the ability to manage inventory, reduce losses, and ensure timely delivery. Fleet management systems utilize IoT to monitor vehicle performance, driver behavior, and fuel consumption, optimizing routes and maintenance schedules to reduce operational costs and environmental impact.

Public transportation systems benefit substantially from IoT applications, offering improved service reliability and passenger experience. Real-time tracking of buses, trains, and other public transport vehicles allows for accurate arrival predictions and better schedule adherence. Passengers can access real-time information through mobile apps, reducing wait times and improving overall satisfaction. Additionally, IoT-based systems can monitor the condition and performance of public transport vehicles, enabling predictive maintenance and minimizing service disruptions.

Furthermore, IoT applications in transportation extend to smart parking solutions, which alleviate the problem of finding available parking spaces in urban areas. Sensors installed in parking lots and on-street parking spaces can detect vehicle presence and communicate availability to drivers via mobile apps. This not only reduces the time and fuel spent searching for parking but also decreases traffic congestion and emissions.

The integration of IoT in transportation also supports the development of autonomous vehicles. These vehicles rely heavily on IoT sensors and communication systems to navigate, make decisions, and interact with their environment. Autonomous vehicles have the potential to significantly reduce accidents caused by human error, optimize traffic flow, and provide mobility solutions for individuals unable to drive.

The deployment of IoT in transportation is not without challenges. Data security and privacy concerns are paramount, as the vast amount of data generated by IoT devices can be susceptible to cyber-attacks. Ensuring interoperability among different IoT systems and devices is also critical to achieving seamless communication and operation. Additionally, the initial investment required for IoT infrastructure can be substantial, though the long-term benefits often justify these costs.

In conclusion, IoT applications in transportation are driving significant advancements in efficiency, safety, and user experience across various domains. As technology continues to evolve, the potential for further innovation and improvement in transportation systems remains vast, promising a more connected and intelligent future.

Case Studies of IoT in Logistics

The integration of the Internet of Things (IoT) into logistics operations has markedly transformed the industry, offering substantial improvements in efficiency, transparency, and cost-effectiveness. This section delves into various case studies illustrating the practical applications and benefits of IoT in logistics.

One notable example is DHL's deployment of IoT technologies to enhance their supply chain operations. By utilizing smart sensors and connected devices, DHL has been able to monitor the real-time location, temperature, and humidity of sensitive shipments. This capability is particularly crucial for the pharmaceutical sector, where maintaining specific conditions is essential for product integrity. The implementation of IoT has enabled DHL to provide clients with real-time data, thereby enhancing trust and reliability. Furthermore, predictive analytics based on IoT data have allowed DHL to optimize routing and reduce the risk of delays, ultimately resulting in cost savings and improved customer satisfaction.

Another significant case is Maersk's use of IoT to revolutionize its container shipping operations. Maersk has equipped its fleet of refrigerated containers, known as "reefers," with IoT sensors that continuously monitor and report on temperature, humidity, and other critical parameters. This monitoring not only ensures the quality of perishable goods but also aids in compliance with stringent international regulations. The data collected from these sensors is transmitted to a centralized platform, enabling Maersk to perform predictive maintenance and avoid potential equipment failures. This proactive approach has led to a reduction in operational downtime and maintenance costs.

The retail giant Amazon has also leveraged IoT to streamline its logistics processes. Through the use of IoT-enabled robotic systems and automated warehouses, Amazon has significantly reduced the time required for order fulfillment. These robots, equipped with sensors and connected to a central system, can efficiently navigate warehouse spaces, locate items, and transport them to human workers for packaging. This integration of IoT has resulted in faster processing times, reduced labor costs, and increased accuracy in order fulfillment. Additionally,

Amazon's use of IoT in delivery vehicles has facilitated real-time tracking and optimization of delivery routes, further enhancing the efficiency of their logistics network.

In the realm of urban logistics, the city of Barcelona has implemented an IoT-based smart waste management system. By equipping waste containers with sensors that monitor fill levels, the city can optimize waste collection routes and schedules. This system has led to a reduction in fuel consumption and carbon emissions, as waste collection trucks only need to service full containers. The data collected also provides insights into waste generation patterns, enabling the city to make informed decisions on resource allocation and urban planning.

IoT has also found applications in enhancing the traceability and security of goods in transit. For instance, IBM and Walmart have collaborated on a blockchain-based system that integrates IoT sensors to track the journey of food products from farm to shelf. This system ensures that each step of the supply chain is transparent and verifiable, thereby reducing the risk of food contamination and fraud. The combination of IoT and blockchain technologies has provided an immutable record of product histories, which is invaluable for quality assurance and regulatory compliance.

These case studies underscore the transformative potential of IoT in logistics. The adoption of IoT technologies has led to significant improvements in operational efficiency, cost reduction, and customer satisfaction. As IoT continues to evolve, its applications in logistics are expected to expand, further driving innovation and excellence in the industry.

Challenges and Future Directions

The rapid evolution of digital technologies has brought about unprecedented opportunities and advancements, yet it also presents a myriad of challenges that must be addressed to ensure sustainable progress. One of the foremost challenges is the digital divide, which continues to widen between those with access to advanced technologies and those without. This disparity not only affects individual opportunities but also has broader societal implications, potentially exacerbating existing inequalities. Bridging this divide requires concerted efforts in policy-making, infrastructure development, and education to ensure that all segments of the population can benefit from digital advancements.

Cybersecurity remains a critical concern as the digital landscape expands. The proliferation of connected devices, from smartphones to Internet of Things (IoT) sensors, increases the potential attack surface for malicious actors. Ensuring robust security measures and developing resilient systems to protect sensitive data and maintain user trust is paramount. This necessitates ongoing research into advanced encryption techniques, threat detection mechanisms, and the development of international cybersecurity standards.

Data privacy is another pressing issue in the digital age. The vast amounts of data generated, collected, and analyzed pose significant risks to individual privacy. The implementation of stringent data protection regulations, such as the General Data Protection Regulation (GDPR) in Europe, marks a step in the right direction, but the challenge remain in balancing the need for data-driven innovation with the protection of personal information. Advances in privacy-preserving technologies, such as differential privacy and federated learning, offer promising avenues to mitigate these concerns.

The rise of artificial intelligence (AI) and machine learning (ML) introduces both opportunities and ethical dilemmas. While these technologies have the potential to revolutionize industries and improve decision-making processes, they also raise questions about accountability, transparency, and bias. Developing fair and explainable AI systems is crucial to prevent the perpetuation of existing biases and to ensure that AI-driven decisions are just and equitable. Interdisciplinary collaboration between technologists, ethicists, and policymakers is essential to address these ethical considerations comprehensively.

Sustainability is another critical challenge in the digital era. The environmental impact of data centers, electronic waste, and the energy consumption associated with digital infrastructure cannot be overlooked. Research into energy-efficient computing, sustainable materials, and recycling processes is necessary to mitigate the environmental footprint of digital technologies. Furthermore, integrating sustainability principles into the design and implementation of digital systems can contribute to more environmentally responsible innovation.

The future direction of the digital highway will be shaped by advancements in emerging technologies such as quantum computing, 5G networks, and Blockchain. Quantum computing holds the promise of solving complex problems that are currently intractable, but it also poses new challenges in terms of security and algorithm design. The deployment of 5G networks is expected to revolutionize connectivity, enabling new applications and services, yet it also requires substantial investment in infrastructure and poses regulatory challenges. Blockchain technology offers potential for enhanced transparency and security in transactions, but its scalability and energy consumption issues need to be addressed.

Addressing these challenges requires a multifaceted approach, involving collaboration between academia, industry, and government. Investment in research and development, coupled with the formulation of forward-thinking policies and regulations, is essential to navigate the complexities of the digital highway. Educating the next generation of technologists and fostering a culture of ethical innovation will be pivotal in shaping a digital future that is inclusive, secure, and sustainable. The journey ahead is fraught with challenges, but the collective efforts of diverse stakeholders can pave the way for a digital landscape that benefits all.

Chapter 3
Artificial Intelligence in Transportation

Introduction to AI

Artificial Intelligence (AI) represents a transformative force in the modern digital landscape, shaping industries, economies, and daily life. This subchapter explores the foundational concepts of AI, tracing its historical development, key principles, and current applications.

The origins of AI can be traced back to the mid-20th century, when pioneering researchers such as Alan Turing and John McCarthy laid the groundwork for what would become a rapidly evolving field. Turing's seminal 1950 paper, "Computing Machinery and Intelligence," posed the question of whether machines could think, introducing the Turing Test as a criterion for machine intelligence. McCarthy, along with Marvin Minsky, Nathaniel Rochester, and Claude Shannon, organized the Dartmouth Conference in 1956, which is widely regarded as the birth of AI as a distinct scientific discipline.

AI encompasses a broad range of subfields, including machine learning, natural language processing, robotics, and computer vision. Machine learning, a core component of AI, involves the development of algorithms that enable computers to learn from and make predictions or decisions based on data. This paradigm shift from rule-based programming to data-driven learning has been pivotal in advancing AI capabilities. Techniques such as supervised learning, unsupervised learning, and reinforcement learning form the backbone of modern machine learning systems.

Natural language processing (NLP) allows machines to understand, interpret, and generate human language. This subfield has seen significant advancements with the advent of deep learning techniques, leading to the development of sophisticated models like OpenAI's GPT-3 and Google's BERT. These models are capable of performing a wide range of language-related tasks, from translation and summarization to sentiment analysis and conversational agents.

Robotics integrates AI with physical machines, enabling autonomous operation in diverse environments. From industrial automation to service robots and autonomous vehicles, the integration of AI in robotics has led to significant improvements in efficiency, precision, and adaptability. Computer vision, another critical area, focuses on enabling machines to interpret and understand visual information from the world. Advances in this field have paved the way for applications ranging from facial recognition and medical imaging to autonomous navigation and surveillance systems.

The impact of AI is evident across various sectors, including healthcare, finance, transportation, and entertainment. In healthcare, AI-driven diagnostic tools and personalized treatment plans are enhancing patient outcomes and operational efficiency. In finance, AI algorithms are used for fraud detection, risk management, and algorithmic trading. Autonomous vehicles, powered by AI, promise to revolutionize transportation by improving safety and reducing congestion. In entertainment, AI is transforming content creation, recommendation systems, and interactive experiences.

Despite its remarkable advancements, AI also presents significant challenges and ethical considerations. Issues such as bias in AI algorithms, data privacy, and the potential for job displacement require careful consideration and regulation. The development of explainable AI, which aims to make AI decision-making processes transparent and understandable, is an important step towards addressing these concerns.

The future of AI holds immense potential, with ongoing research and innovation poised to further expand its capabilities and applications. As AI continues to evolve, it will undoubtedly play a crucial role in shaping the digital highway, driving progress and innovation in the years to come.

AI Applications in Road Logistics

Advancements in artificial intelligence (AI) have significantly impacted the road logistics sector, offering transformative solutions that enhance operational efficiency, safety, and cost-effectiveness. AI-driven applications in road logistics can be broadly categorized into route optimization, predictive maintenance, autonomous vehicles, and demand forecasting.

Route optimization stands out as one of the most impactful AI applications in road logistics. Traditional route planning methods rely heavily on static data and manual inputs, making them susceptible to inefficiencies and errors. In contrast, AI algorithms can process vast amounts of real-time data, including traffic conditions, weather forecasts, and historical route performance, to dynamically adjust routes. These algorithms leverage machine learning techniques to predict potential delays and suggest alternative paths, thereby reducing travel time and fuel consumption. Companies such as UPS and DHL have already integrated AI-based route optimization tools into their logistics operations, resulting in substantial cost savings and improved delivery times.

Predictive maintenance is another critical area where AI is making significant inroads. The traditional approach to vehicle maintenance is often reactive, addressing issues only after they arise. This can lead to unexpected breakdowns and costly downtime. AI systems, equipped with sensors and data analytics capabilities, can monitor the condition of vehicle components in real-time. Machine learning models analyze this data to predict potential failures before they occur, allowing for timely maintenance interventions. Predictive maintenance not only extends the lifespan of vehicles but also enhances overall fleet reliability and safety.

The development of autonomous vehicles represents a revolutionary shift in road logistics. Self-driving trucks, powered by sophisticated AI systems, promise to address several challenges, including driver shortages, regulatory compliance, and operational inefficiencies. These vehicles utilize a combination of sensors, cameras, and LiDAR technology to navigate roads safely and efficiently. LiDAR involves sending laser light from a transmitter and reflecting it off objects in the scene. The system receiver detects the reflected light, and the TOF is utilized to create a distance map of the objects in the scene. AI algorithms process sensory data to make real-time driving decisions, such as lane changes, speed adjustments, and obstacle avoidance. Companies like Waymo and Tesla are at the forefront of developing autonomous trucks, conducting extensive testing to ensure safety and reliability. The widespread adoption of autonomous vehicles could potentially reduce labor costs and improve delivery precision, although regulatory and ethical considerations remain significant hurdles.

Demand forecasting is essential for optimizing inventory levels and ensuring timely deliveries. AI algorithms can analyze historical sales data, market trends, and external factors such as economic indicators and seasonal variations to predict future demand with high accuracy. These predictive models enable logistics companies to make informed decisions about inventory management, reducing the risk of stockouts and overstocking. Enhanced demand forecasting capabilities also facilitate better resource allocation, as companies can adjust their logistics operations to align with anticipated demand fluctuations.

AI applications in road logistics also extend to enhancing customer experience. AI-powered chatbots and virtual assistants provide real-time updates on shipment status, address customer inquiries, and resolve issues efficiently. These tools improve customer satisfaction by offering timely and accurate information, reducing the need for human intervention.

The integration of AI in road logistics is not without challenges. Data privacy and security concerns, the need for substantial investment in technology infrastructure, and the requirement for skilled personnel to manage and interpret AI systems are significant considerations. However, the potential benefits of AI applications in road logistics far outweigh these challenges, promising a future of more efficient, reliable, and cost-effective logistics operations.

In conclusion, AI is poised to revolutionize road logistics through its applications in route optimization, predictive maintenance, autonomous vehicles, and demand forecasting. As technology continues to evolve, the logistics sector must adapt to leverage these advancements, ensuring a competitive edge in an increasingly digital landscape.

Case Studies of AI in Transportation

Artificial Intelligence (AI) is increasingly becoming a pivotal component in the transportation sector, offering innovative solutions that enhance efficiency, safety, and sustainability. This subchapter explores various case studies that exemplify the transformative impact of AI in transportation, focusing on real-world applications and their outcomes.

One prominent example is the deployment of AI in autonomous vehicles. Companies like Tesla and Waymo have pioneered the integration of machine learning algorithms and sensor technologies to enable self-driving capabilities. These systems utilize a combination of cameras, LiDAR, and radar to perceive the environment, while neural networks process this data to make real-time driving decisions. The results have been promising, with significant reductions in human error-related accidents and improved traffic flow. However, challenges remain, particularly in complex urban environments where unpredictable factors can still disrupt autonomous systems.

Another notable case is the use of AI in predictive maintenance for public transportation systems. The Massachusetts Bay Transportation Authority (MBTA) has implemented machine learning models to analyze data from sensors embedded in trains and tracks. These models predict potential failures before they occur, allowing for timely maintenance and reducing downtime. The initiative has led to a marked improvement in service reliability and cost savings, demonstrating the potential of AI to enhance operational efficiency.

AI is also making strides in traffic management. The city of Hangzhou in China has adopted an AI-driven traffic control system developed by Alibaba Cloud. This system collects and analyzes data from an extensive network of cameras and sensors across the city. By leveraging deep learning algorithms, it optimizes traffic signal timings and provides real-time traffic predictions. The implementation has resulted in a significant reduction in traffic congestion and travel time, showcasing the capability of AI to manage complex urban traffic systems effectively.

In the realm of logistics, AI is revolutionizing supply chain management. DHL has incorporated AI algorithms to optimize route planning and delivery schedules. By analyzing vast amounts of data, including weather conditions, traffic patterns, and delivery urgency, the system can dynamically adjust routes to ensure timely deliveries. This has not only enhanced operational efficiency but also reduced fuel consumption and carbon emissions, aligning with sustainability goals.

AI's role in enhancing passenger experience is also noteworthy. Airlines such as Delta and Emirates are employing AI-powered chatbot's and virtual assistants to provide personalized customer service. These systems utilize natural language processing (NLP) to understand and respond to passenger queries, offering real-time information on flight status, baggage tracking, and more. The adoption of AI in customer service has led to improved passenger satisfaction and operational efficiency.

The integration of AI in maritime transportation is another compelling case. Ports like the Port of Rotterdam have implemented AI systems to optimize cargo handling and vessel traffic management. Machine learning algorithms analyze data from various sources, including ship movements, weather forecasts, and tidal patterns, to predict optimal docking times and streamline port operations. This has resulted in increased throughput and reduced waiting times for vessels, demonstrating the efficacy of AI in managing complex maritime logistics.

These case studies highlight the diverse applications of AI in the transportation sector, each contributing to enhanced efficiency, safety, and sustainability. The successful implementation of AI technologies underscores their potential to address longstanding challenges and drive innovation in transportation. As AI continues to evolve, its integration into various facets of transportation is expected to deepen, offering new opportunities for optimization and improvement.

Ethical and Practical Considerations

The rapid advancement of digital technologies has brought about a plethora of opportunities and challenges. As we navigate this digital highway, it becomes imperative to consider both ethical and practical aspects to ensure responsible and sustainable development. The integration of digital systems into various facets of daily life necessitates a thorough examination of the implications for privacy, security, and equity.

Foremost among the ethical considerations is the issue of privacy. The collection, storage, and analysis of vast amounts of personal data have raised significant concerns. The potential for misuse of this data by corporations, governments, or malicious actors underscores the need for stringent data protection measures. Regulatory frameworks such as the General Data Protection Regulation (GDPR) in the European Union set a precedent for safeguarding individual privacy. However, the global nature of digital interactions demands harmonized international standards to effectively address these concerns.

Security is another critical aspect that intertwines with ethical considerations. The proliferation of cyber threats, ranging from data breaches to sophisticated ransomware attacks, necessitates robust security protocols. Ethical hacking, or penetration testing, has emerged as a valuable practice to identify and mitigate vulnerabilities within digital systems. Organizations must prioritize cybersecurity to protect sensitive information and maintain trust with users. Furthermore, the ethical implications of surveillance technologies, such as facial recognition, must be carefully weighed against their potential benefits in areas like law enforcement and public safety.

Equity and access to digital technologies present another layer of ethical complexity. The digital divide, characterized by disparities in access to technology and the internet, exacerbates existing social inequalities. Efforts to bridge this divide should focus on providing affordable and reliable internet access, particularly in underserved communities. Additionally, digital literacy programs are essential to empower individuals with the skills needed to navigate and benefit from digital technologies. Ensuring equitable access to digital resources is not only a matter of social justice but also a practical necessity for fostering inclusive economic growth.

From a practical standpoint, the integration of digital technologies into various sectors requires careful planning and implementation. Interoperability between different systems and platforms is crucial to maximize efficiency and effectiveness. Standardization of protocols and data formats can facilitate seamless communication and collaboration across diverse digital ecosystems. Moreover, the scalability and sustainability of digital solutions must be considered to avoid obsolescence and ensure long-term viability.

The workforce implications of digital transformation also warrant attention. Automation and artificial intelligence (AI) have the potential to disrupt labor markets, leading to job displacement in certain sectors. It is essential to proactively address these changes through reskilling and upskilling initiatives. Collaborative efforts between governments, educational institutions, and industry stakeholders can help workers transition to new roles and adapt to the evolving demands of the digital economy.

Ethical considerations extend to the environmental impact of digital technologies. The energy consumption of data centers and the electronic waste generated by obsolete devices contribute to the environmental footprint of the digital landscape. Sustainable practices, such as the adoption of energy-efficient technologies and the promotion of the circular economy, are vital to mitigate these effects. Policymakers and industry leaders must prioritize sustainability to balance technological advancement with environmental stewardship.

In conclusion, the ethical and practical considerations surrounding digital technologies are multifaceted and interdependent. Addressing these issues requires a holistic and collaborative approach that balances innovation with responsibility. By prioritizing privacy, security, equity, interoperability, workforce adaptation, and sustainability, we can navigate the digital highway in a manner that benefits society as a whole.

Chapter 4

Blockchain Technology in Supply Chain Management

Understanding Blockchain

Blockchain technology represents a revolutionary paradigm in the realm of digital information management, offering unprecedented security, transparency, and decentralization. At its core, Blockchain is a distributed ledger technology (DLT) that records transactions across multiple computers so that the registered transactions cannot be altered retroactively. This immutability is achieved through cryptographic hashing and consensus mechanisms, which collectively ensure the integrity and chronological order of data entries.

The fundamental structure of a Blockchain consists of a series of blocks, each containing a list of transactions. Each block is linked to the previous one through a cryptographic hash, forming a continuous chain. This linkage not only ensures the immutability of the data but also provides a verifiable history of all transactions. A critical component of this system is the consensus algorithm, which governs how new blocks are added to the chain. Common consensus mechanisms include Proof of Work (PoW), Proof of Stake (PoS), and Practical

Byzantine Fault Tolerance (PBFT), each with its own method of validating transactions and maintaining network security.

The decentralized nature of Blockchain technology eliminates the need for a central authority, thereby reducing the risk of single points of failure and enhancing the resilience of the network. This decentralization is achieved through a peer-to-peer (P2P) network, where each participant, or node, maintains a copy of the entire Blockchain. This redundancy ensures that even if multiple nodes fail or act maliciously, the overall integrity of the Blockchain remains intact. Furthermore, the use of cryptographic techniques ensures that only authorized parties can initiate transactions, providing a robust security framework.

Blockchain technology is not limited to financial transactions; its applications extend to various sectors such as supply chain management, healthcare, and voting systems. In supply chain management, for instance, Blockchain can enhance transparency and traceability by providing an immutable record of the journey of goods from production to consumption. In healthcare, Blockchain can secure patient records, ensuring that they are accessible only to authorized personnel while maintaining a comprehensive and tamper-proof history of medical data. Voting systems can leverage Blockchain to create transparent and verifiable election processes, reducing the potential for fraud and increasing public trust.

Smart contracts are another significant innovation enabled by blockchain technology. These are self-executing contracts with the terms of the agreement directly written into code. They automatically enforce and execute the terms of the contract when predefined conditions are met, eliminating the need for intermediaries and reducing the possibility of human error or manipulation. Platforms like Ethereum have popularized smart contracts, enabling a multitude of decentralized applications (dApps) that operate on blockchain networks.

The scalability of blockchain technology remains a critical challenge. As the number of transactions increases, the time and computational power required to validate and add new blocks also rise. Various solutions, such as sharding and off-chain transactions, are being explored to address these limitations. Sharding involves partitioning the blockchain into smaller, more manageable pieces, while off-chain transactions allow for certain transactions to be processed outside the main blockchain, reducing the load on the primary network.

Blockchain technology continues to evolve, with ongoing research and development aimed at enhancing its capabilities and addressing its limitations. Its potential to transform various industries and redefine the way digital information is managed makes it a pivotal component of the digital highway.

Blockchain Use Cases in Logistics

Blockchain technology, originally developed as the underlying infrastructure for crypto currencies, has found a significant application in the field of logistics. The decentralized and immutable nature of Blockchain provides a robust framework for enhancing transparency, security, and efficiency in logistics operations. This section explores the various use cases of Blockchain technology within the logistics sector, demonstrating its transformative potential.

One of the primary applications of Blockchain in logistics is in the area of supply chain visibility. Traditional supply chains often suffer from a lack of transparency, leading to inefficiencies and increased risk of fraud. Blockchain addresses these issues by providing a single, immutable ledger that records all transactions and movements of goods. Each participant in the supply chain can access this ledger, ensuring that all parties have a consistent and real-time view of the status and location of goods. This enhanced visibility not only reduces the risk of fraud but also allows for more accurate demand forecasting and inventory management.

Another critical application is in the area of traceability. In industries such as food and pharmaceuticals, the ability to trace the origin and journey of products is crucial for ensuring quality and safety. Blockchain enables end-to-end traceability by recording every transaction and movement of products on a tamper-proof ledger. This ensures that in the event of a recall or quality issue, companies can quickly and accurately trace the affected products back to their source, thereby reducing the risk of harm to consumers and minimizing financial losses.

Smart contracts are another innovative application of Blockchain in logistics. These are self-executing contracts with the terms of the agreement directly written into code. Smart contracts can automate various processes within the logistics chain, such as payments and customs clearance. For instance, a smart contract could be programmed to automatically release payment to a supplier, once a shipment has been delivered and verified. This reduces the need for intermediaries, speeds up transactions, and reduces the risk of human error.

Blockchain also enhances security in logistics operations. Traditional logistics networks are vulnerable to cyberattacks and data breaches, which can lead to significant financial losses and reputational damage. The decentralized nature of Blockchain makes it inherently more secure, as there is no central point of failure. Additionally, the cryptographic techniques used in Blockchain ensure that data is tamper-proof and cannot be altered without the consensus of the network participants. This makes Blockchain an ideal solution for securing sensitive information, such as shipment details and transaction records.

The application of Blockchain in logistics is not without its challenges. One of the primary obstacles is the need for widespread adoption and standardization across the industry. For blockchain to be truly effective, all participants in the supply chain must adopt the technology and adhere to common standards. This requires significant investment in technology and training, as well as collaboration between different stakeholders. Additionally, there are regulatory and legal considerations that must be addressed, particularly concerning data privacy and cross-border transactions.

Despite these challenges, the potential benefits of Blockchain in logistics are substantial. By providing greater transparency, traceability, and security, Blockchain can significantly enhance the efficiency and reliability of logistics operations. As the technology continues to evolve and mature, it is likely that its adoption in the logistics sector will continue to grow, paving the way for a more efficient and a secure global supply chain.

Benefits of Blockchain in Supply Chains

Blockchain technology has garnered significant attention in recent years, particularly for its potential to revolutionize supply chain management. Traditional supply chain systems are often plagued by inefficiencies, lack of transparency, and susceptibility to fraud. Blockchain offers a decentralized, immutable ledger that can address these issues, providing a more efficient, transparent, and secure supply chain.

One of the primary benefits of Blockchain in supply chains is enhanced transparency. In traditional supply chain systems, information is often siloed, making it difficult to track the movement of goods and verify their authenticity. Blockchain enables all participants in the supply chain to access a single, immutable ledger that records every transaction. This transparency allows for real-time tracking of goods, reducing the risk of fraud and errors. For instance, in the food industry, Blockchain can be used to track the journey of a product from farm to table, ensuring that it meets safety standards and reducing the likelihood of contamination.

Another significant advantage is increased efficiency. Traditional supply chains often involve multiple intermediaries, each adding time and cost to the process. Blockchain can streamline these operations by automating many of the tasks currently performed by intermediaries. Smart contracts, which are self-executing contracts with the terms directly written into code, can be used to automate payments, shipments, and other transactions. This automation reduces the need for manual intervention, speeding up the process and reducing costs. For example, in the automotive industry, Blockchain can be used to automate the procurement of parts, ensuring that they are delivered just in time for production, thereby reducing inventory costs.

Blockchain also enhances security in supply chains. Traditional systems are vulnerable to cyber-attacks and fraud, as they often rely on a central database that can be compromised. Blockchain's decentralized nature makes it much more difficult for a single point of failure to occur. Each participant in the supply chain has a copy of the ledger, and any changes must be validated by the network. This makes it extremely difficult for malicious actors to alter the data. Additionally, the use of cryptographic techniques ensures that data is secure and can only be accessed by authorized parties.

Moreover, blockchain can improve trust and collaboration among supply chain participants. In traditional systems, trust is often established through intermediaries or third-party audits, which can be time-consuming and costly. Blockchain provides a trustless environment where transactions are verified by the network, eliminating the need for intermediaries. This can foster greater collaboration among participants, as they can rely on the accuracy and integrity of the data recorded on the Blockchain. For example, in the pharmaceutical industry, Blockchain can be used to ensure that drugs are sourced from legitimate suppliers and have not been tampered with, thereby improving trust between manufacturers, distributors, and consumers.

Additionally, blockchain can facilitate regulatory compliance. Supply chains are often subject to a myriad of regulations, and ensuring compliance can be a complex and costly process. Blockchain provides a transparent and an immutable record of transactions, making it easier to demonstrate compliance with regulatory requirements. For instance, in the logistics sector, blockchain can be used to track the movement of goods across borders, ensuring that they comply with customs regulations and reducing the risk of delays.

A prime example of this is **Farm to Plate**, a US-based company, that has successfully integrated blockchain technology into its logistics operations, to address complex regulatory and tracking challenges. Farm to Plate specializes in the farm-to-fork supply chain, ensuring that agricultural products are tracked from their origin on the farm to the final consumer's plate. By utilizing blockchain, they maintain an unbroken, transparent record of the journey of each product, ensuring compliance with food safety standards, reducing waste, and increasing overall efficiency.

Farm to Plate's blockchain solution not only ensures regulatory compliance but also builds trust with consumers, by providing them with verifiable information about the origins and journey of their food. This approach is revolutionizing, how logistics and supply chain management are handled in the agricultural sector, and it serves as a model for other industries facing similar challenges.

By embracing blockchain technology, companies like Farm to Plate are not only solving logistical challenges but are also setting new standards for transparency, efficiency, and regulatory compliance in the global supply chain.

In sum, blockchain technology offers numerous benefits for supply chain management, including enhanced transparency, increased efficiency, improved security, greater trust and collaboration, and facilitated regulatory compliance. These advantages make blockchain a promising solution for addressing the challenges faced by traditional supply chain systems and paving the way for more efficient, secure, and transparent supply chains.

Regulatory and Implementation Challenges

The rapid advancement of digital technologies has necessitated a comprehensive framework to manage and regulate their implementation. A primary challenge lies in the inherently global nature of digital technologies, which often transcend national borders and regulatory jurisdictions. This creates a complex landscape where harmonizing international regulations becomes essential, yet exceedingly difficult. Divergent national policies can lead to fragmentation, impeding the seamless integration of digital systems and services.

Data privacy and security remain paramount concerns. The General Data Protection Regulation (GDPR) in the European Union sets a high bar for data protection, influencing policies worldwide. However, discrepancies in data privacy laws across different regions can create compliance challenges for multinational corporations. Ensuring that data is managed in accordance with local regulations, while maintaining a cohesive global strategy requires, significant resources and sophisticated technological solutions.

Interoperability is another critical issue. As various sectors adopt digital technologies, the lack of standardized protocols can hinder the integration of systems. For instance, in the healthcare industry, disparate electronic health record (EHR) systems can obstruct the seamless exchange of patient information, thereby affecting the quality of care. Developing and enforcing universal standards is crucial to facilitate interoperability, yet achieving consensus among diverse stakeholders is a formidable task.

The deployment of digital infrastructure also encounters logistical and financial barriers. Rural and underserved areas often lack the necessary investments for high-speed internet and advanced digital services. Bridging this digital divide requires coordinated efforts between governments, private sector entities, and international organizations. Public-private partnerships can play a pivotal role in mobilizing resources and expertise to expand digital infrastructure to these regions.

Cybersecurity threats pose a significant risk to the implementation of digital technologies. As systems become more interconnected, they also become more vulnerable to cyber-attacks. Establishing robust cybersecurity frameworks and ensuring continuous monitoring and updating of these systems is critical. This requires not only technological solutions but also regulatory measures that mandate stringent security protocols and facilitate rapid response mechanisms.

Moreover, the ethical implications of digital technologies, such as artificial intelligence (AI) and machine learning, necessitate thoughtful regulation. Issues like algorithmic bias, transparency, and accountability must be addressed to ensure that these technologies are deployed in a fair and equitable manner. Regulatory bodies must work closely with technologists, ethicists, and civil society to develop guidelines that safeguard against misuse while fostering innovation.

The pace of technological change often outstrips the ability of regulatory frameworks to adapt. This lag can result in outdated or inadequate regulations that fail to address new challenges. Regulatory bodies need to adopt more agile approaches, incorporating mechanisms for regular review and updates to keep pace with technological advancements. This may involve the use of regulatory sandboxes, where new technologies can be tested in a controlled environment before broader implementation.

In conclusion, balancing innovation with regulation is a delicate act that requires ongoing dialogue and collaboration among all stakeholders. Policymakers must navigate the intricate web of technical, ethical, and logistical challenges to create an environment where digital technologies can thrive while safeguarding public interest. Effective regulation and implementation strategies are essential to harness the full potential of the digital highway, ensuring that its benefits are equitably distributed and its risks are mitigated.

Chapter 5

Autonomous Vehicles and Their Impact

The Rise of Autonomous Vehicles

The advent of autonomous vehicles represents a significant milestone in the evolution of transportation technology. This development is rooted in advances in artificial intelligence, sensor technology, and data analytics, which have collectively enabled the creation of vehicles capable of navigating complex environments without human intervention. Autonomous vehicles, or self-driving vehicles, are equipped with an array of sensors, including LiDAR, radar, and cameras, which provide a comprehensive understanding of the surrounding environment. These sensors generate vast amounts of data, which are processed in real-time by sophisticated algorithms to make driving decisions.

The concept of autonomous vehicles dates back to the mid-20th century, but substantial progress has been made in the past two decades. Early prototypes were limited by the computational power and sensor technology available at the time. However, recent advancements in machine learning and neural networks have significantly improved the ability of these systems to interpret sensor data and learn from experience. This has led to the development of more reliable and efficient autonomous driving systems.

One of the key components of autonomous vehicles is the perception system, which enables the vehicle to detect and identify objects, obstacles, and road conditions. LiDAR, which stands for Light Detection and Ranging, plays a crucial role in this system by providing high-resolution 3D maps of the vehicle's surroundings. Radar complements LiDAR by offering robust detection capabilities in adverse weather conditions, while cameras provide detailed visual information. The fusion of data from these sensors allows the vehicle to construct an accurate and dynamic model of its environment.

Another critical aspect is the decision-making system, which involves path planning and control algorithms. Path planning algorithms determine the optimal route for the vehicle to follow, taking into account factors such as traffic, road conditions, and destination. Control algorithms, on the other hand, execute the planned path by adjusting the vehicle's speed, steering, and braking. These algorithms must operate with high precision and reliability to ensure the safety and efficiency of the autonomous vehicle.

The integration of vehicle-to-everything (V2X) communication technology further enhances the capabilities of autonomous vehicles. V2X communication allows vehicles to exchange information with other vehicles, infrastructure, and pedestrians, providing additional layers of situational awareness. This connectivity can improve traffic flow, reduce accidents, and enable cooperative driving strategies.

Despite the remarkable progress, several challenges remain in the widespread adoption of autonomous vehicles. Ensuring the safety and reliability of these systems is paramount, as even minor errors can have significant consequences. Regulatory frameworks and standards need to be established to govern the deployment and operation of autonomous vehicles. Additionally, public acceptance and trust in this technology must be cultivated through education and transparent communication about its benefits and limitations.

The potential benefits of autonomous vehicles are substantial. They promise to reduce traffic accidents, alleviate congestion, and provide mobility solutions for individuals who are unable to drive. Moreover, autonomous vehicles could lead to more efficient use of resources, such as fuel and road space, contributing to environmental sustainability.

The rise of autonomous vehicles is a testament to the transformative power of digital technology. As research and development continue to advance, the realization of fully autonomous transportation systems is becoming increasingly feasible. This technological evolution holds the promise of reshaping the future of mobility, with far-reaching implications for society, the economy, and the environment.

Impact on Road Logistics

The advent of digital technologies has significantly altered the landscape of road logistics, a critical component of the global supply chain. One of the primary ways in which digitalization has impacted road logistics is through enhanced route optimization. Advanced algorithms and real-time data analytics enable logistics companies to determine the most efficient routes for their fleets. This not only reduces fuel consumption but also minimizes delivery times, thereby improving overall operational efficiency. GPS technology, coupled with IoT sensors, provides real-time tracking of vehicles, allowing for dynamic rerouting in response to traffic conditions, accidents, or other unforeseen events.

Another significant impact of digitalization in road logistics is the improvement in fleet management. Fleet management systems now incorporate telematics to monitor various aspects of vehicle performance, such as fuel efficiency, engine health, and driver behavior.

Predictive maintenance, powered by machine learning algorithms, helps in anticipating and addressing potential vehicle issues before they result in breakdowns. This proactive approach reduces downtime and maintenance costs, enhancing the reliability of logistics operations.

Digital platforms have revolutionized the way freight is matched with available transport capacity. Online freight marketplaces and load boards provide a platform for shippers and carriers to connect, enabling more efficient utilization of transport resources. This has led to a reduction in empty miles, where trucks travel without cargo, thus optimizing fuel usage and reducing carbon emissions. Furthermore, Blockchain technology is being explored to enhance transparency and security in the logistics chain, providing immutable records of transactions and movements, which is particularly beneficial for high-value or sensitive shipments.

The integration of digital technologies has also facilitated better communication and coordination among various stakeholders in the logistics ecosystem. Cloud-based logistics management systems allow for seamless information sharing between shippers, carriers, and customers. This enhanced visibility leads to better coordination and planning, reducing the likelihood of delays and improving customer satisfaction. Moreover, automated notifications and updates keep all parties informed about shipment status, further enhancing transparency.

The rise of autonomous vehicles and advanced driver-assistance systems (ADAS) represents another frontier in the digital transformation of road logistics. Autonomous trucks, equipped with sophisticated sensors and AI, promise to revolutionize long-haul transport by reducing the need for human drivers. This could address the chronic shortage of truck drivers and reduce labor costs. ADAS technologies, such as adaptive cruise control, lane-keeping assist, and collision avoidance systems, enhance the safety and efficiency of both autonomous and human-driven vehicles.

E-commerce growth has placed additional demands on road logistics, necessitating more frequent and smaller deliveries. Digital solutions, such as last-mile delivery optimization algorithms and drone delivery, are being explored to meet these demands. Last-mile delivery is often the most expensive and complex part of the logistics chain, but digital tools help in optimizing delivery routes, reducing costs, and improving delivery times.

Regulatory compliance and environmental sustainability are also being addressed through digitalization. Electronic logging devices (ELDs) ensure compliance with hours-of-service regulations, enhancing driver safety and reducing fatigue-related accidents. Digital tools for emissions monitoring help logistics companies adhere to environmental regulations and work towards sustainability goals.

The transformation brought about by digital technologies in road logistics is multifaceted, encompassing route optimization, fleet management, freight matching, stakeholder communication, autonomous vehicles, last-mile delivery, and regulatory compliance. These advancements not only improve operational efficiency and safety but also contribute to sustainability and customer satisfaction, fundamentally reshaping the road logistics landscape.

Case Studies of Autonomous Vehicles

The advent of autonomous vehicles (AVs) marks a significant milestone in the evolution of transportation systems. These sophisticated machines, leveraging advanced sensors, machine learning algorithms, and real-time data processing, promise to redefine mobility paradigms. To elucidate the practical implications and technological prowess of AVs, we examine several case studies that highlight their deployment and performance in varied contexts.

One of the most prominent examples is Waymo, the autonomous vehicle subsidiary of Alphabet Inc. Waymo has conducted extensive testing and deployment in different urban environments, most notably in Phoenix, Arizona. Utilizing a combination of LiDAR, radar, and camera systems, Waymo's vehicles have accumulated millions of miles in both simulated and real-world conditions. The Phoenix pilot program, which includes a ride-hailing service, demonstrates the potential for AVs to operate safely and efficiently in a bustling city landscape. Waymo's rigorous safety protocols and continuous data collection underscore the importance of iterative learning and adaptation in the development of autonomous driving technologies.

Another significant case study is Tesla's deployment of its Full Self-Driving (FSD) system. Unlike Waymo, Tesla opts for a vision-based approach using cameras and neural networks, eschewing LiDAR. Tesla's FSD system is integrated into its consumer vehicles, enabling features such as autonomous lane-changing, parking, and navigation on highways. Tesla's approach emphasizes the incremental deployment of autonomous features through over-the-air updates, allowing the company to collect vast amounts of data from its fleet. This data-driven strategy facilitates continuous improvement and real-time responsiveness to diverse driving conditions. However, Tesla's approach has sparked debates on the readiness and safety of deploying partially autonomous systems to the general public.

In Europe, the collaboration between the UK government and Oxbotica, an autonomous vehicle software company, provides another illustrative example. Oxbotica has been involved in several projects, including the DRIVEN consortium, which aimed to deploy a fleet of autonomous vehicles in Oxford and London. The focus on urban environments with complex traffic scenarios and pedestrian interactions provides valuable insights into the challenges and capabilities of AVs in densely populated areas. Oxbotica's emphasis on modular and scalable software solutions highlights the need for flexibility and adaptability in AV technologies, catering to various vehicle types and operational domains.

The use of autonomous vehicles in logistics and freight transportation is exemplified by the Swedish company Einride. Einride's autonomous electric trucks, known as T-pods, are designed for short-haul freight operations. These vehicles operate in controlled environments such as warehouses and industrial parks, where they can navigate predefined routes with minimal human intervention. The integration of electric propulsion systems with autonomous driving technology showcases the potential for AVs to contribute to sustainable and efficient logistics solutions. Einride's deployments demonstrate significant reductions in both operational costs and environmental impact, aligning with broader goals of reducing carbon emissions and enhancing supply chain efficiency.

These case studies collectively illustrate the diverse applications and technological strategies employed in the development and deployment of autonomous vehicles. While each example presents unique challenges and achievements, common themes such as safety, data-driven development, and adaptability emerge as critical factors in the advancement of AV technologies. The ongoing evolution of autonomous vehicles continues to shape the future of transportation, offering promising solutions to contemporary mobility challenges.

Safety and Regulatory Issues

The rapid advancement of digital technologies has led to widespread adoption across various sectors, from healthcare to transportation. However, as these technologies become more integrated into daily life, they bring with them a host of safety and regulatory challenges that must be addressed to ensure their responsible and secure use. This subchapter delves into the critical aspects of safety and regulatory issues associated with the digital highway.

One of the primary concerns is cybersecurity. With the increasing reliance on interconnected systems, the potential for cyber-attacks has grown exponentially.

These attacks can lead to significant disruptions, data breaches, and even physical harm if critical infrastructures such as power grids or transportation systems are compromised. Ensuring robust cybersecurity measures is paramount. This includes implementing advanced encryption techniques, regularly updating software to patch vulnerabilities, and conducting continuous monitoring to detect and mitigate threats promptly.

Data privacy is another significant issue. The digital highway generates vast amounts of data, much of which is personal and sensitive. Regulatory frameworks like the General Data Protection Regulation (GDPR) in Europe have been established to protect individuals' privacy rights. These regulations mandate stringent requirements for data collection, storage, and usage, compelling organizations to adopt transparent data handling practices. Compliance with such regulations is not only a legal obligation but also a crucial step in building public trust in digital technologies.

The proliferation of autonomous systems, particularly in the automotive sector, introduces new safety considerations. Autonomous vehicles (AVs) rely on complex algorithms and real-time data processing to navigate and make decisions. Ensuring the reliability of these systems is critical to prevent accidents and ensure passenger safety. This necessitates rigorous testing and validation protocols, along with the development of standards that define acceptable performance levels for AVs. Regulatory bodies must work closely with industry stakeholders to establish these standards and ensure that

AVs meet stringent safety criteria before they are deployed on public roads.

Artificial intelligence (AI) and machine learning (ML) technologies, which underpin many digital innovations, also pose unique regulatory challenges. These technologies have the potential to transform industries, but they also raise concerns about bias, accountability, and transparency. Regulatory frameworks must be developed to address these issues, ensuring that AI and ML systems are designed and deployed in a manner that is fair, transparent, and accountable. This includes establishing guidelines for the ethical use of AI, mandating regular audits to detect and mitigate biases, and requiring clear documentation of decision-making processes.

The Internet of Things (IoT) further complicates the regulatory landscape. IoT devices, ranging from smart home appliances to industrial sensors, are often designed with limited security features, making them vulnerable to cyber-attacks. Regulatory standards must be established to ensure that IoT devices are secure by design. This includes requiring manufacturers to implement robust security measures, such as encryption and secure boot processes, and mandating regular software updates to address emerging threats.

The digital highway also intersects with existing regulatory frameworks in various domains, such as telecommunications, finance, and healthcare. Harmonizing these regulations to accommodate the unique challenges posed by digital technologies is essential. This involves updating existing laws to reflect the new realities of the digital age and ensuring that regulatory bodies have the expertise and resources needed to effectively oversee the deployment and use of digital technologies.

Addressing safety and regulatory issues is a complex but necessary task to ensure that the digital highway can be navigated safely and responsibly. By establishing robust cybersecurity measures, protecting data privacy, ensuring the reliability of autonomous systems, developing ethical AI and ML frameworks, securing IoT devices, and harmonizing existing regulations, society can harness the benefits of digital technologies while mitigating associated risks.

Chapter 6
Enhancing Efficiency in Road Logistics

Technological Innovations Driving Efficiency

The rapid advancement of technology has significantly transformed various sectors, enhancing efficiency and productivity. Notably, the integration of digital tools and systems has revolutionized traditional processes, leading to unprecedented improvements in operational performance. Several technological innovations stand out for their impact on efficiency, including automation, artificial intelligence (AI), and the Internet of Things (IoT).

Automation has become a cornerstone in achieving operational efficiency. Through the use of robotics and automated systems, repetitive and mundane tasks are executed with precision and speed far surpassing human capabilities. For instance, in manufacturing, robotic arms and automated assembly lines have drastically reduced the time required for production while minimizing errors. This level of precision and consistency is unattainable through manual labor, thus underscoring the critical role of automation in modern industries.

Artificial intelligence further amplifies efficiency by enabling predictive analytics and decision-making processes. Machine learning algorithms analyze vast amounts of data to identify patterns and predict outcomes, thereby facilitating proactive measures. In the healthcare sector, AI-driven diagnostic tools can rapidly analyze medical images and patient data to detect anomalies, leading to early intervention and improved patient outcomes. Similarly, in finance, AI algorithms assess market trends and risks, enabling more informed investment decisions. The ability to process and interpret large datasets with high accuracy accelerates decision-making and optimizes resource allocation.

The Internet of Things (IoT) represents another significant leap toward enhanced efficiency. By interconnecting devices and systems, IoT creates a network of smart objects that communicate and collaborate seamlessly. In logistics, IoT-enabled sensors track the location and condition of goods in real-time, ensuring timely deliveries and reducing losses. Smart grids in the energy sector monitor and manage electricity distribution, optimizing energy consumption and reducing waste. The integration of IoT in various applications underscores its potential to streamline operations and improve overall efficiency.

Moreover, cloud computing has emerged as a pivotal technology in driving efficiency. By providing scalable and on-demand computing resources, cloud services eliminate the need for extensive physical infrastructure. This not only reduces costs but also enhances flexibility and accessibility. Organizations can rapidly deploy applications and services, leveraging the cloud's computational power to handle complex tasks. For instance, data-intensive research projects benefit from cloud-based high-performance computing, enabling faster analysis and discoveries. The ability to access and share resources globally fosters collaboration and accelerates innovation.

Blockchain technology also contributes to operational efficiency by ensuring transparency and security in transactions. The decentralized nature of blockchain eliminates intermediaries, reducing transaction times and costs. In supply chain management, blockchain provides a tamper-proof ledger that records every transaction, enhancing traceability and accountability. This level of transparency is crucial in sectors like pharmaceuticals, where the authenticity of products is paramount. By streamlining processes and securing data, blockchain technology enhances efficiency across various domains.

The convergence of these technological innovations creates a synergistic effect, further amplifying their individual benefits. For example, the integration of AI with IoT devices enables smarter and more responsive systems. Predictive maintenance powered by AI algorithms and IoT sensors can foresee equipment failures, allowing timely interventions and minimizing downtime. Such interconnected systems epitomize the future of efficient operations, where technology continuously adapts and optimizes processes.

In conclusion, the ongoing advancements in technology continue to drive efficiency across multiple sectors. Automation, AI, IoT, cloud computing, and blockchain are pivotal in this transformation, each contributing uniquely to enhanced operational performance. The synergy between these technologies promises a future where efficiency is maximized, and productivity is elevated to new heights.

Optimizing Supply Chain Management

Supply chain management has undergone substantial transformation with the advent of digital technologies, leading to enhanced efficiency, reduced costs, and improved responsiveness. Digital tools such as Artificial Intelligence (AI), Internet of Things (IoT), and Blockchain have revolutionized traditional supply chain processes. The integration of these technologies facilitates real-time data analysis, predictive analytics, and seamless communication across various stakeholders, thereby optimizing the entire supply chain network.

Artificial Intelligence plays a pivotal role in supply chain optimization by enabling advanced data analytics and decision-making capabilities. AI algorithms can process vast amounts of data to forecast demand accurately, optimize inventory levels, and identify potential disruptions. Machine learning models, a subset of AI, continuously learn from historical data to improve predictions and adapt to changing market conditions. This predictive capability allows companies to anticipate demand fluctuations and adjust their production schedules accordingly, minimizing both overstocking and stockouts.

The Internet of Things enhances supply chain visibility by connecting physical assets to the digital world. IoT devices such as sensors and RFID tags collect real-time data on the location, condition, and status of goods. This data is transmitted to centralized platforms where it can be analyzed to monitor the movement of products through the supply chain. Enhanced visibility enables proactive management of potential issues such as delays, damages, or theft, thereby ensuring the timely delivery of goods. Additionally, IoT facilitates the automation of routine tasks, such as inventory tracking and reordering, reducing manual intervention and errors.

Blockchain technology introduces a new level of transparency and security to supply chain management. By providing a decentralized and immutable ledger, blockchain ensures that every transaction is recorded and cannot be altered. This transparency builds trust among supply chain partners and enhances traceability, enabling companies to verify the authenticity and origin of products. Blockchain can also streamline administrative processes by automating contract execution through smart contracts, which are self-executing contracts with the terms of the agreement directly written into code.

The implementation of digital twins further exemplifies the transformative potential of digital technologies in supply chain optimization. Digital twins are virtual replicas of physical assets, processes, or systems that can be used to simulate and analyze real-world scenarios. By creating digital twins of their supply chains, companies can run simulations to identify bottlenecks, evaluate the impact of potential changes, and optimize overall performance. This capability allows for better contingency planning and more informed decision-making.

Integration of these technologies requires a robust digital infrastructure and a strategic approach. Companies must invest in the necessary hardware and software, and train their workforce to leverage these tools effectively. Data security and privacy are also critical considerations, as the increased flow of information across digital platforms can expose companies to cyber threats. Therefore, implementing stringent cybersecurity measures is essential to protect sensitive data and maintain the integrity of the supply chain.

The convergence of AI, IoT, blockchain, and Digital Twins is driving a paradigm shift in supply chain management. These technologies collectively enhance visibility, predictability, and efficiency, leading to more resilient and agile supply chains. As companies continue to adopt and integrate these digital tools, the future of supply chain management promises to be more interconnected, transparent, and responsive to the dynamic demands of the global market.

Real-Time Tracking and Monitoring

Real-time tracking and monitoring have emerged as pivotal components in the digital infrastructure, leveraging advanced technologies to provide instantaneous data acquisition and analysis. The integration of Internet of Things (IoT) devices, Geographic Information Systems (GIS), and sophisticated data analytics platforms has revolutionized various sectors, including logistics, healthcare, and environmental monitoring. This subchapter delves into the mechanisms, applications, and implications of real-time tracking and monitoring within the digital landscape.

At the core of real-time tracking systems are IoT devices equipped with sensors and communication modules. These devices continuously collect data on various parameters such as location, temperature, and movement. The data is transmitted to central servers via cellular networks, satellite communication, or Wi-Fi, ensuring seamless connectivity and data flow. The integration of GPS technology further enhances the accuracy and reliability of location-based data, facilitating precise tracking of assets and individuals.

In logistics, real-time tracking systems enable companies to monitor the movement of goods throughout the supply chain. By utilizing RFID tags and GPS-enabled devices, businesses can achieve end-to-end visibility of their inventory, thus optimizing route planning and reducing transit times. This real-time insight into the supply chain not only improves operational efficiency but also enhances customer satisfaction by providing accurate delivery timelines.

In a recent interview, **Mr. Viral Shah, Executive Director of V-Trans (India) Ltd**, emphasized the critical role of technology in transforming logistics operations. He stated, "At V-Trans, we believe that integrating advanced tracking systems is not just about keeping pace with industry trends; it's about setting new standards in operational excellence and customer service. Real-time visibility is a game-changer, allowing us to proactively manage our supply chain and respond swiftly to any challenges. This commitment to innovation is what drives us to continuously improve our services and exceed customer expectations." Mr. Shah's insights reflect V-Trans' dedication to leveraging cutting-edge technology to stay ahead in the competitive logistics landscape, ensuring that every shipment is delivered with precision and reliability.

Real-time monitoring is a crucial component of driver support within the transportation industry. Sensors integrated into wearable devices can constantly monitor vitals including, but not limited to, heart rate, blood pressure and fatigue levels. This real time information is then relayed to fleet managers so any health issues can be acted upon at speed. In taking control of the well-being of your drivers, not only are these safety barriers at their maximum potential but when a crash does happen and even in trying to prevent it from happening being more efficient than anyone else is simply one call away.

The integration of Artificial Intelligence (AI) and Machine Learning (ML) algorithms with real-time tracking systems further amplifies their capabilities. These technologies can analyze vast amounts of data in real-time, identifying patterns and predicting future events. In the context of smart cities, AI-powered real-time monitoring can optimize traffic flow, reduce energy consumption, and enhance public safety. By analyzing data from various sensors, AI can predict traffic congestion and suggest alternative routes, thereby minimizing delays and improving urban mobility.

Despite the numerous advantages, real-time tracking and monitoring systems also pose significant challenges. Data privacy and security are paramount concerns, as the continuous transmission of sensitive information can be susceptible to cyber threats. Ensuring robust encryption and secure data storage is crucial to safeguarding user data. Additionally, the deployment of extensive sensor networks requires substantial investment in infrastructure and maintenance, which can be a limiting factor for widespread adoption.

Ethical considerations also arise from the pervasive nature of real-time monitoring. The potential for surveillance and the intrusion of privacy necessitates stringent regulatory frameworks to balance the benefits of real-time tracking with individual rights. Transparent data governance policies and the development of privacy-preserving technologies are essential to address these concerns.

Real-time tracking and monitoring represent a significant advancement in the digital highway, offering unparalleled insights and capabilities across various sectors. The continued evolution and integration of these systems promise to drive further innovation, enhancing efficiency, safety, and sustainability in an increasingly connected world.

Reducing Operational Costs

The advent of digital technologies has precipitated a paradigm shift in how businesses manage and reduce operational costs. Through the integration of advanced digital tools and processes, organizations can streamline operations, enhance efficiency, and ultimately achieve significant cost savings. This chapter delves into the multifaceted approaches enabled by digital transformation that contributes to reduced operational expenditures.

One of the primary mechanisms by which digital technologies reduce costs is through automation. Automation of routine tasks, whether in manufacturing, in logistics, or in administrative functions, minimizes the need for manual intervention and reduces labor costs. For instance, robotic process automation (RPA) allows for the execution of repetitive tasks with precision and speed, freeing up human resources for more complex and strategic activities.

The COVID-19 pandemic brought back memories of my leadership role in spearheading the successful implementation of the electronic Lorry Receipt (eLR) project, which set a new precedent in the Indian transportation industry. Lorry Receipts were traditionally created by the operations team on the ground through a manual process. The pandemic caused unprecedented difficulties, with lockdowns and restrictions leaving no one to handle these vital tasks. Our team recognized the urgency and created a fully automated eLR system, which is the first of its kind in India's transport history.

The timing of this project was crucial as it enabled seamless documentation and ensured continuity in transportation operations when it was most needed. The system we developed allows for automated generation of eLRs, with data being fetched directly from the eWay Bill portal to ensure accuracy and compliance. This innovation demonstrated to the industry that such critical documents could be prepared digitally and sent to the consignee, consignor, and truckers with digital signatures, paving the way for a more efficient and resilient future in logistics.

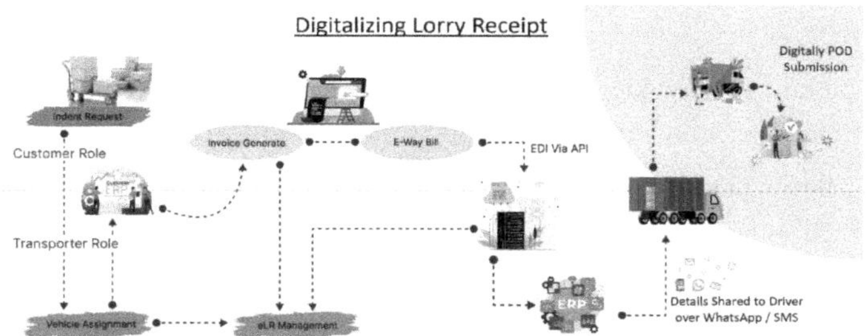

Another critical aspect of digital transformation is the optimization of supply chain management. Digital technologies such as the Internet of Things (IoT), blockchain, and advanced analytics provide real-time visibility into supply chain operations. This enhanced visibility enables organizations to identify inefficiencies, predict demand more accurately, and manage inventory levels effectively. The use of IoT sensors, for example, allows for real-time monitoring of equipment and inventory, reducing the likelihood of overstocking or stockouts. Blockchain technology offers a secure and transparent way to track goods throughout the supply chain, reducing fraud and errors and ensuring compliance with regulatory requirements.

Cloud computing is another pivotal element in reducing operational costs. By migrating to cloud-based platforms, organizations can significantly reduce on the expenses associated with maintaining physical IT infrastructure.

Cloud services offer scalable solutions that can be adjusted based on the organization's needs, thus avoiding the costs of over-provisioning and underutilization of resources. Moreover, it facilitates remote work, reducing the need for physical office space and associated overhead costs.

Data analytics and artificial intelligence (AI) play a crucial role in cost reduction by enabling more informed decision-making. Predictive analytics can forecast market trends, customer behavior, and operational bottlenecks, allowing organizations to proactively address potential issues before they escalate into costly problems. AI-driven tools can optimize various aspects of operations, from energy consumption in data centers to dynamic pricing strategies in retail, ensuring that resources are used efficiently and cost-effectively.

Furthermore, digital transformation fosters a culture of continuous improvement. By leveraging digital tools for performance monitoring and feedback, organizations can identify areas for improvement and implement changes swiftly. This agility reduces the time and resources spent on rectifying issues and enhances overall operational efficiency. For example, digital twins—virtual replicas of physical assets—allow for real-time monitoring and simulation, enabling predictive maintenance and reducing downtime.

In addition to these direct cost-saving measures, digital technologies also contribute to indirect cost reductions. Improved customer service through digital channels can enhance customer satisfaction and loyalty, reducing the costs associated with acquiring and retaining customers.

Enhanced cybersecurity measures protect against data breaches and cyberattacks, which can be financially devastating.

Overall, the strategic implementation of digital technologies offers a comprehensive approach to reducing operational costs. By automating processes, optimizing supply chains, leveraging cloud computing, utilizing data analytics and AI, fostering continuous improvement, and enhancing customer service and cybersecurity, organizations can achieve substantial cost efficiencies. The integration of these digital solutions not only reduces expenses but also positions businesses for sustained growth and competitiveness in the digital era.

Chapter 7
Improving Safety with Technology

Safety Challenges in Road Logistics

Road logistics, integral to global commerce, face myriad safety challenges that demand rigorous attention. The increasing volume of goods transported by road, coupled with the complexity of logistics networks, introduces significant risks that necessitate robust safety measures. The primary safety concerns in road logistics encompass vehicular accidents, cargo security, and environmental hazards.

Vehicular accidents remain a predominant safety issue. The frequency and severity of road incidents involving logistics vehicles are influenced by factors such as driver fatigue, vehicle maintenance, and road conditions. Driver fatigue, often resulting from long hours and tight delivery schedules, impairs reaction times and decision-making abilities, thereby elevating the risk of accidents. Regular vehicle maintenance is critical to ensuring roadworthiness and minimizing mechanical failures that could lead to accidents. Additionally, road conditions, including weather and infrastructure quality, play a crucial role in accident prevention. Advanced driver-assistance systems (ADAS) and autonomous driving technologies are being developed to mitigate these risks, yet their widespread adoption faces technological and regulatory hurdles.

Cargo security is another pivotal concern in road logistics. The high value of transported goods makes them attractive targets for theft and pilferage. Ensuring the security of cargo involves implementing measures such as tamper-evident seals, GPS tracking, and surveillance systems. However, these measures are not foolproof and can be circumvented by sophisticated methods of theft. The integration of blockchain technology is being explored to enhance transparency and traceability in the supply chain, potentially reducing the incidence of cargo theft. Furthermore, training and awareness programs for drivers and logistics personnel are essential to recognize and respond to security threats effectively.

Environmental hazards also pose significant safety challenges in road logistics. The transportation of hazardous materials, such as chemicals and flammable substances, requires stringent safety protocols to prevent accidents and environmental concerns. Compliance with regulations governing the handling and transportation of hazardous materials is imperative to mitigate risks. In addition, the logistics industry must contend with the broader environmental impact of road transportation, including emissions and fuel consumption. The adoption of greener technologies, such as electric and hydrogen-powered vehicles, is being pursued to address these environmental concerns, though the transition faces economic and infrastructural challenges.

The human element in road logistics safety cannot be overlooked. Driver training and performance significantly influence safety outcomes. Comprehensive training programs that cover defensive driving techniques, fatigue management, and emergency response are crucial for enhancing driver competency and reducing accident rates. The implementation of telematics and monitoring systems can provide real-time data on driver performance, enabling targeted interventions to improve safety.

Regulatory frameworks play a vital role in shaping safety standards in road logistics. Compliance with national and international regulations ensures a baseline level of safety, but there is often variability in enforcement and adherence. Harmonizing regulations across different jurisdictions can facilitate safer logistics operations and promote best practices.

The advent of digital technologies presents opportunities to address these safety challenges more effectively. IoT devices, big data analytics, and machine learning can enhance predictive maintenance, optimize route planning, and improve real-time monitoring of logistics operations. However, the successful integration of these technologies requires overcoming barriers related to data privacy, interoperability, and cybersecurity.

In conclusion, addressing the safety challenges in road logistics necessitates a multifaceted approach that combines technological innovation, regulatory compliance, and human factors management. The continuous evolution of digital technologies offers promising avenues for enhancing safety, yet their effective implementation demands coordinated efforts from all stakeholders in the logistics ecosystem.

Technological Solutions for Safety

The rapid advancement of digital technologies has significantly influenced the transportation sector, particularly in enhancing safety measures along the digital highway. A myriad of technological solutions has been developed and implemented to mitigate risks and ensure the protection of all road users. These solutions can be categorized into several key areas: vehicle-based technologies, infrastructure enhancements, and communication systems.

Vehicle-based technologies have seen significant innovation with the integration of advanced driver-assistance systems (ADAS). These systems utilize a combination of sensors, cameras, and artificial intelligence to assist drivers in making safer decisions. Features such as automatic emergency braking (AEB), lane-keeping assistance (LKA), and adaptive cruise control (ACC) have been proven to reduce the likelihood of collisions. Additionally, the development of autonomous vehicles, equipped with sophisticated algorithms and real-time data processing capabilities, promises to further diminish human error, which is a leading cause of accidents.

Infrastructure enhancements play a crucial role in bolstering safety on the digital highway. Intelligent transportation systems (ITS) have been widely adopted to manage traffic flow and reduce congestion, thereby decreasing the potential for accidents. These systems include smart traffic signals that adjust in real-time based on traffic conditions, dynamic message signs that provide drivers with up-to-date information, and automated incident detection systems that enable quicker emergency response times. The implementation of these technologies has resulted in more efficient and safer transportation networks.

Communication systems form the backbone of many safety technologies. Vehicle-to-everything (V2X) communication allows vehicles to exchange information with each other and with infrastructure elements. This exchange enables the anticipation of potential hazards and coordination of responses to dynamic traffic conditions. For instance, V2X technology can alert drivers of upcoming road work, sudden braking by vehicles ahead, or adverse weather conditions. By facilitating this real-time communication, V2X significantly enhances situational awareness and response time.

One notable application of V2X is the development of cooperative adaptive cruise control (C-ACC), which extends the capabilities of traditional adaptive cruise control by enabling vehicles to communicate and synchronize their movements. This cooperation reduces the likelihood of collisions and improves traffic flow efficiency. Furthermore, the integration of 5G technology is expected to enhance the reliability and speed of V2X communication, providing an even more robust framework for safety applications.

In addition to these technological advancements, data analytics and machine learning have become integral in predicting and preventing accidents. By analyzing vast amounts of data collected from various sources, such as traffic cameras, sensors, and historical accident records, predictive models can identify high-risk areas and times. These insights allow for proactive measures, such as targeted enforcement and infrastructure improvements, to be implemented.

The ongoing development and deployment of these technological solutions is crucial for achieving safer transportation systems. As these technologies continue to evolve and integrate, they hold the potential to drastically reduce the incidence of accidents and fatalities on the digital highway. The collaboration between vehicle manufacturers, technology developers, and regulatory bodies will be essential in ensuring these innovations are effectively and universally applied.

Case Studies of Safety Improvements

The integration of digital technologies into transportation systems has yielded significant advancements in safety. Various case studies illustrate how specific implementations have led to measurable improvements, providing valuable insights for future applications.

One notable example is the deployment of Advanced Driver Assistance Systems (ADAS) in urban environments. ADAS technologies, including features such as adaptive cruise control, lane-keeping assistance, and automated emergency braking, have been instrumental in reducing collision rates. A study conducted in a major metropolitan area demonstrated a 22% reduction in rear-end collisions and a 17% decrease in side-impact crashes following the widespread adoption of ADAS-equipped vehicles. This data underscores the efficacy of these systems in mitigating common types of vehicular accidents.

Another significant case is the application of Vehicle-to-Everything (V2X) communication technologies. V2X enables vehicles to communicate with each other and with infrastructure, such as traffic lights and road sensors. In a pilot project in a mid-sized city, V2X technology was implemented at several high-risk intersections. The results were compelling: incidents of intersection-related accidents dropped by 30%, and near-miss occurrences were reduced by 45%. The ability of V2X to provide real-time information and alerts to drivers played a crucial role in these improvements, highlighting its potential to enhance situational awareness and prevent accidents.

Interview with Mr. Harvinder Singh Banga, CIO, CJ Darcl:

Mr. Harvinder Singh Banga, CIO of CJ Darcl, shared his thoughts on the impact of advanced technologies on road safety. "At CJ Darcl, we are exploring ADAS and V2X communication technologies into our fleet, which has significantly enhanced safety and reduced incidents caused by driver error. V2X, in particular, has improved our ability to manage logistics in urban areas, enabling real-time communication between our vehicles and traffic signals."

He added, "These technologies not only help in preventing accidents but also contribute to a safer and more efficient transportation system. Our focus remains on adopting innovations that enhance both safety and operational efficiency."

The use of predictive analytics in traffic management centers represents another transformative safety improvement. By leveraging big data and machine learning algorithms, traffic management systems can predict and respond to potential hazards more effectively. In a large-scale experiment conducted on a busy highway corridor, predictive analytics were used to anticipate traffic congestion and reroute vehicles accordingly. This approach resulted in a 15% reduction in traffic-related accidents and a 20% improvement in emergency response times. The predictive models allowed for proactive measures, demonstrating the value of data-driven decision-making in traffic safety.

Additionally, the implementation of smart infrastructure has shown promising results. Smart roadways equipped with sensors and IoT devices can monitor and report real-time conditions such as weather, traffic flow, and road surface status. In a coastal region prone to fog and slippery conditions, smart infrastructure was installed along a critical highway segment. The system provided timely alerts to drivers and traffic management authorities about hazardous conditions. Subsequent evaluations revealed a 28% decrease in weather-related accidents and a 35% improvement in road safety during adverse conditions. The real-time data provided by smart infrastructure allowed for immediate and informed responses, significantly enhancing driver safety.

Furthermore, the integration of digital technologies in public transportation systems has also contributed to safety improvements. The introduction of automated monitoring and control systems in buses and trains has reduced human error and enhanced operational safety. In a case study involving a commuter rail network, the implementation of an automated train control system led to a 40% reduction in operational incidents and a 25% decrease in passenger injuries. The system's ability to maintain optimal speeds, enforce safety protocols, and provide real-time diagnostics was pivotal in achieving these outcomes.

These case studies collectively illustrate the profound impact of digital technologies on transportation safety. Each implementation demonstrates a unique application of technology, yet all share a common goal: reducing accidents and enhancing the overall safety of transportation systems. The successes observed in these cases provide a roadmap for further advancements, underscoring the critical role of digital innovation in creating safer highways and transit networks.

Future Trends in Safety Technology

The rapid development of technology continues to reshape various aspects of our lives, and the domain of safety technology is no exception. Emerging trends indicate a future where digital advancements will significantly enhance safety measures across multiple sectors, including transportation, manufacturing, and personal security. This subchapter delves into the pivotal trends likely to dominate the safety technology landscape in the foreseeable future.

One of the most impactful trends is the integration of artificial intelligence (AI) and machine learning (ML) into safety systems. AI and ML algorithms can analyze vast amounts of data in real-time, identifying potential hazards and predicting accidents before they occur. For instance, in the automotive industry, advanced driver-assistance systems (ADAS) employ AI to monitor driving conditions, detect obstacles, and make split-second decisions to prevent collisions. Similarly, in industrial settings, predictive maintenance powered by AI can anticipate equipment failures, thereby reducing the risk of accidents and ensuring operational continuity.

The Internet of Things (IoT) is another transformative trend that promises to revolutionize safety technology.

IoT devices, embedded with sensors and connected to the internet, can collect and transmit data continuously. In smart cities, IoT-enabled infrastructure can monitor environmental conditions, such as air quality and traffic flow, to enhance public safety. Wearable IoT devices, such as smart helmets and vests, are being developed for construction workers to monitor vital signs and environmental conditions, providing real-time alerts in case of hazardous situations.

Blockchain technology, primarily known for its applications in finance, is also making inroads into safety technology. Blockchain's decentralized and immutable nature ensures the integrity and security of data, which is crucial for safety-critical applications. For example, in the supply chain industry, blockchain can track and verify the authenticity of products, reducing the risk of counterfeit goods that could pose safety hazards. Additionally, blockchain can be used to secure communication between autonomous vehicles, ensuring that data exchanged is tamper-proof and reliable.

Augmented reality (AR) and virtual reality (VR) are emerging as powerful tools for safety training and simulation. AR and VR technologies create immersive environments where users can experience realistic scenarios without the associated risks. In aviation, VR simulators are used for pilot training, allowing them to practice emergency procedures in a controlled environment. In healthcare, AR can assist surgeons by overlaying critical information onto the patient's body, enhancing precision and reducing the likelihood of errors.

Cybersecurity is becoming increasingly critical as safety technology becomes more interconnected. Protecting these systems from cyber threats is paramount to ensuring their reliability and effectiveness. Advances in cybersecurity measures, such as advanced encryption techniques and intrusion detection systems, are essential to safeguard the integrity of safety technologies. Ensuring robust cybersecurity protocols will mitigate the risk of malicious attacks that could compromise safety systems.

5G technology is set to play a significant role in the future of safety technology by providing high-speed, low-latency communication. This capability is crucial for applications requiring real-time data transmission, such as autonomous vehicles and remote medical procedures. The enhanced connectivity offered by 5G will enable more efficient and effective safety solutions, facilitating quicker response times and more accurate data analysis.

The convergence of these technological trends points to a future where safety measures are more proactive, predictive, and interconnected. The ongoing advancements in AI, IoT, blockchain, AR/VR, cybersecurity, and 5G will collectively forge a new era of safety technology, characterized by unprecedented levels of efficiency and reliability. It is certain that these technologies will play a crucial role in safeguarding human lives and assets in an increasingly digital world as they evolve.

Chapter 8

Environmental Impact and Sustainability

Environmental Challenges in Logistics

The logistics sector plays a pivotal role in the global economy, facilitating the movement of goods from producers to consumers. However, this essential function comes with significant environmental challenges that need to be addressed to ensure sustainable development. One of the primary environmental concerns in logistics is the emission of greenhouse gases (GHGs). The transportation of goods, particularly via road and air, is a major contributor to carbon dioxide (CO2) emissions. These emissions contribute to global warming and climate change, posing a serious threat to the environment.

In addition to GHG emissions, logistics operations often result in air pollution. The combustion of fossil fuels in vehicles releases pollutants such as nitrogen oxides (NOx) and particulate matter (PM), which have adverse effects on air quality and public health. Urban areas, where logistics activities are concentrated, are particularly susceptible to the detrimental impacts of air pollution.

The logistics industry also significantly impacts land use and biodiversity. The construction of infrastructure such as warehouses, distribution centers, and transportation networks often leads to habitat destruction and fragmentation. This can result in the loss of biodiversity and disruption of ecosystems. Moreover, the expansion of logistics infrastructure can contribute to urban sprawl, further exacerbating environmental degradation.

Waste generation is another critical environmental challenge in logistics. Packaging materials, including plastics, cardboard, and other disposables, contribute to the growing problem of waste management. Improper disposal of these materials can lead to pollution of land and water bodies, harming wildlife and ecosystems. Additionally, the logistics industry generates electronic waste (e-waste) from the use of technology and equipment, which requires proper handling and recycling to mitigate its environmental impact.

Energy consumption in logistics operations is a significant concern. The sector relies heavily on non-renewable energy sources, particularly fossil fuels, for transportation and warehousing. This dependence on fossil fuels not only contributes to GHG emissions but also depletes finite natural resources. Renewable energy alternatives and energy-efficient technologies are essential to reduce the environmental footprint of logistics activities.

The environmental challenges in logistics are compounded by the increasing demand for rapid delivery and the growth of e-commerce. The rise of same-day and next-day delivery services has intensified the pressure on logistics networks, leading to increased vehicle miles traveled (VMT) and higher emissions. The proliferation of e-commerce has also resulted in more frequent and smaller shipments, which are less efficient and more environmentally taxing.

To address these environmental challenges, the logistics industry must adopt sustainable practices and innovative solutions. This includes the use of alternative fuels such as electric and hydrogen-powered vehicles, which produce lower emissions compared to traditional internal combustion engines. The implementation of green logistics strategies, such as optimizing transportation routes and consolidating shipments, can enhance efficiency and reduce environmental impact.

Advancements in digital technologies offer significant potential to mitigate environmental challenges in logistics. The use of big data analytics, Internet of Things (IoT) devices, and artificial intelligence (AI) can improve supply chain visibility and enable more efficient resource management. For example, predictive analytics can optimize inventory levels and reduce waste, while IoT sensors can monitor vehicle performance and enhance fuel efficiency.

Collaboration among stakeholders, including governments, businesses, and consumers, is crucial to drive the transition towards sustainable logistics. Regulatory frameworks and incentives can encourage the adoption of environmentally friendly practices, while consumer awareness and demand for green products can further propel the industry towards sustainability.

The logistics sector must navigate these environmental challenges to achieve a balance between economic growth and environmental stewardship. By leveraging technology and embracing sustainable practices, the industry can contribute to a more sustainable future while continuing to fulfill its vital role in the global economy.

Sustainable Technologies and Practices

The advent of digital technologies has brought about significant advancements in various fields, but it has also raised concerns regarding environmental sustainability. The integration of sustainable technologies and practices within the digital realm is imperative to mitigate the ecological footprint of these innovations. This subchapter examines the pivotal role of sustainable technologies and practices in the digital highway, emphasizing their importance and implementation.

Energy consumption is a critical concern in the digital age. Data centers, which are the backbone of digital infrastructure, consume vast amounts of energy. Implementing energy-efficient technologies in these centers is essential. Techniques such as virtualization, which allows multiple virtual servers to run on a single physical server, can significantly reduce energy consumption. Additionally, the adoption of advanced cooling systems, such as liquid cooling and free cooling, can further enhance energy efficiency by reducing the need for air conditioning.

Renewable energy sources are another vital component of sustainable digital practices. Solar, wind, and hydroelectric power can be harnessed to provide clean energy for digital operations. Companies like Google and Apple have already made substantial investments in renewable energy, powering their data centers with 100% renewable energy. This shift not only reduces carbon emissions but also sets a precedent for other organizations to follow.

The concept of a circular economy is gaining traction in the digital sector. This approach focuses on minimizing waste and maximizing the lifecycle of products. In the context of digital technologies, this can be achieved through practices such as refurbishing and recycling electronic devices. Extending the life of electronic gadgets through repair and upgrades can significantly reduce electronic waste. Moreover, recycling programs that recover valuable materials from discarded electronics help conserve natural resources and reduce environmental impact.

Green software engineering is another emerging field that aims to develop software with minimal energy consumption. This involves optimizing algorithms and code to reduce the computational power required, thereby lowering energy usage. Efficient software design can lead to substantial energy savings, especially when deployed at scale in data centers and cloud computing environments.

The deployment of Internet of Things (IoT) technologies presents both challenges and opportunities for sustainability. While IoT devices can lead to increased energy consumption due to the proliferation of connected devices, they also offer potential for significant energy savings through smart management systems. For instance, smart grids can optimize electricity distribution, reducing energy waste. Similarly, smart buildings equipped with IoT sensors can enhance energy efficiency by automatically adjusting lighting, heating, and cooling systems based on occupancy and usage patterns.

Blockchain technology, often criticized for its high energy consumption, is also evolving towards more sustainable practices. Proof-of-stake (PoS) consensus mechanisms, for example, offer a more energy-efficient alternative to the traditional proof-of-work (PoW) systems used by cryptocurrencies like Bitcoin. PoS significantly reduces the computational power required for transaction validation, thereby lowering energy consumption.

Sustainable practices in digital manufacturing are crucial as well. Additive manufacturing, commonly known as 3D printing, can reduce waste by using only the necessary materials for production. This technology also allows for localized manufacturing, which can decrease the carbon footprint associated with transportation and logistics.

The integration of sustainable technologies and practices within the digital highway is not merely an option but a necessity. As the digital landscape continues to expand, the adoption of energy-efficient technologies, renewable energy sources, circular economy principles, green software engineering, and sustainable IoT and blockchain practices will play a crucial role in reducing the environmental impact. The commitment to sustainability in the digital age is essential for ensuring a balanced coexistence between technological progress and environmental stewardship.

Case Studies of Sustainable Logistics

The integration of digital technologies into logistics has paved the way for more sustainable practices across various industries. This chapter examines three case studies that highlight the successful implementation of digital solutions to achieve sustainability goals. Each case demonstrates the potential for significant environmental and economic benefits through the adoption of innovative logistics strategies.

The first case study focuses on a multinational retail corporation that implemented an advanced telematics system to optimize its fleet management. By utilizing GPS tracking, real-time data analytics, and predictive maintenance algorithms, the company was able to significantly reduce fuel consumption and greenhouse gas emissions. The telematics system provided insights into driver behavior, allowing for targeted training programs that promoted fuel-efficient driving techniques. Additionally, predictive maintenance reduced vehicle downtime, ensuring that the fleet operated at optimal efficiency. The results included a 15% reduction in fuel usage and a 10% decrease in overall operational costs, demonstrating the efficacy of digital tools in fostering sustainable logistics.

The second case study examines a global shipping company that leveraged blockchain technology to enhance transparency and efficiency in its supply chain. The blockchain-based platform enabled secure and immutable tracking of goods from origin to destination, reducing the risk of frauds and errors. This increased transparency allowed for better coordination among stakeholders, leading to more efficient routing and scheduling. The immutable nature of blockchain records also facilitated compliance with environmental regulations, ensuring that all parties adhered to sustainability standards. The implementation of this technology resulted in a 20% reduction in administrative costs and a 12% decrease in carbon emissions, highlighting the potential of blockchain to drive sustainable logistics practices.

The third case study explores the use of artificial intelligence (AI) and machine learning (ML) by an e-commerce giant to optimize its warehousing and distribution processes. By analyzing vast amounts of data, AI algorithms were able to predict demand patterns with high accuracy, allowing for more efficient inventory management. This predictive capability minimized overstocking and understocking, reducing waste and improving resource utilization. Furthermore, ML models optimized delivery routes based on real-time traffic data, weather conditions, and other variables, leading to faster and more fuel-efficient deliveries. The adoption of AI and ML technologies resulted in a 25% improvement in delivery times and a 15% reduction in energy consumption, underscoring the transformative impact of digital innovation on sustainable logistics.

These case studies illustrate the diverse ways in which digital technologies can be harnessed to achieve sustainability in logistics. The implementation of telematics, blockchain, and AI/ML technologies not only enhances operational efficiency but also contributes to significant environmental benefits. As industries continue to face increasing pressure to adopt sustainable practices, these examples provide valuable insights into the potential of digital solutions to meet and exceed sustainability targets. The lessons learned from these case studies can serve as a blueprint for other organizations seeking to integrate digital technologies into their logistics operations, driving both economic and environmental gains.

Future Directions in Sustainability

As digital technologies continue to evolve, their integration into sustainability initiatives presents both opportunities and challenges. The convergence of digital innovation with sustainable practices has the potential to significantly mitigate environmental impact and promote resource efficiency. This subchapter explores emerging trends and future directions in the application of digital technologies to enhance sustainability.

One promising area is the Internet of Things (IoT), which facilitates real-time monitoring and management of resources. IoT-enabled sensors and devices can optimize energy consumption, reduce waste, and improve operational efficiency across various sectors. For instance, smart grids leverage IoT to balance energy supply and demand dynamically, reducing reliance on fossil fuels and enhancing the integration of renewable energy sources. Similarly, in agriculture, IoT devices can monitor soil conditions and weather patterns, enabling precision farming techniques that minimize water usage and chemical inputs.

Artificial intelligence (AI) and machine learning (ML) also hold significant potential for advancing sustainability. AI algorithms can analyze vast datasets to identify patterns and predict outcomes, aiding in the development of more efficient systems. In the context of urban planning, AI can optimize traffic flow, reducing congestion and lowering emissions. Furthermore, AI-driven predictive maintenance can extend the lifespan of machinery and infrastructure, decreasing the need for resource-intensive replacements.

Blockchain technology offers another avenue for promoting sustainability. Its decentralized and transparent nature can enhance supply chain traceability, ensuring that products are sourced and produced sustainably. Blockchain can verify the authenticity of certifications, such as organic or fair-trade labels, and track the carbon footprint of goods throughout their lifecycle. This level of transparency can drive consumer awareness and demand for sustainable products, incentivizing companies to adopt greener practices.

The role of Digital Twins should also be considered in the sustainability landscape. Digital Twins are virtual replicas of physical systems that can simulate and predict the behavior of their real-world counterparts. These models can optimize the design and operation of systems, from manufacturing processes to urban infrastructure, leading to significant resource savings. For example, Digital Twins of buildings can simulate energy usage patterns, informing retrofits that enhance energy efficiency and reduce greenhouse gas emissions.

The integration of digital technologies with circular economy principles is another critical direction. The circular economy model emphasizes the reuse, recycling, and remanufacturing of products to create closed-loop systems that minimize waste. Digital platforms can facilitate the sharing and refurbishment of goods, extending their lifecycle and reducing the need for new resource extraction. Advanced data analytics can also identify opportunities for material recovery and recycling, optimizing resource flows within industrial ecosystems.

However, the proliferation of digital technologies is not without its challenges. The environmental footprint of digital infrastructure, including data centers and electronic waste, must be addressed. Data centers consume substantial amounts of energy, necessitating innovations in energy-efficient computing and the adoption of renewable energy sources. Additionally, the lifecycle management of electronic devices, from production to disposal, requires sustainable practices to mitigate e-waste.

The intersection of digital innovation and sustainability is a dynamic and rapidly evolving field. Interdisciplinary collaboration among technologists, policymakers, businesses, and researchers is essential to harness the full potential of digital technologies for sustainable development. By prioritizing sustainability in the design and deployment of digital solutions, it is possible to create systems that not only drive economic growth but also preserve and enhance the natural environment for future generations.

Chapter 9
Data Analytics and Big Data

Role of Data Analytics in Logistics

Data analytics has become an integral component in the logistics sector, transforming how operations are managed and decisions are made. With the exponential growth of data generated by various sources, logistics companies are increasingly leveraging advanced analytics techniques to enhance efficiency, reduce costs, and improve customer satisfaction.

The vast amount of data available in logistics includes information from supply chain activities, transportation networks, inventory levels, and customer interactions. Analyzing this data allows companies to gain valuable insights into their operations. Predictive analytics, for instance, enables firms to forecast demand more accurately, optimizing inventory levels and reducing the risk of overstocking or stock outs. This ensures that products are available when and where they are needed, increasing the reliability of the supply chain.

Moreover, data analytics facilitates route optimization in transportation management. By analyzing historical traffic patterns, weather conditions, and real-time data, logistics firms can determine the most efficient routes for their fleets. This not only reduces fuel consumption and operational costs but also minimizes delivery times, enhancing customer satisfaction. Additionally, advanced analytics can help in identifying potential disruptions in the supply chain, allowing companies to take proactive measures to mitigate risks.

The application of machine learning algorithms in logistics is another area, where data analytics is making a significant impact. These algorithms can analyze vast datasets to identify patterns and anomalies that may not be apparent through traditional analysis methods. For instance, machine learning can predict equipment failures in warehouses or transportation vehicles, enabling preventive maintenance and reducing downtime. This leads to more reliable operations and lower maintenance costs.

Data analytics also plays a crucial role in enhancing the transparency and visibility of the supply chain. By integrating data from various sources, companies can create a comprehensive view of their supply chain activities. This visibility allows for better coordination among different stakeholders, including suppliers, manufacturers, and retailers. Enhanced transparency helps in identifying bottlenecks and inefficiencies, enabling companies to address them promptly and improve overall supply chain performance.

In addition to operational improvements, data analytics contributes to strategic decision-making in logistics. By analyzing market trends, customer preferences, and competitive dynamics, companies can make informed decisions about expanding into new markets, launching new products, or adjusting pricing strategies. This strategic use of data analytics provides a competitive edge in a rapidly evolving market environment.

Furthermore, data analytics supports sustainability initiatives in logistics. By optimizing routes, reducing idle times, and improving load planning, companies can lower their carbon footprint and contribute to environmental conservation. Analytics can also help in monitoring and managing energy consumption in warehouses and distribution centers, promoting more sustainable practices.

The integration of data analytics in logistics is not without challenges. Data quality and consistency are critical for accurate analysis, and companies must invest in robust data management systems. Additionally, the complexity of supply chain networks requires sophisticated analytical models and skilled personnel to interpret the results effectively. Despite these challenges, the benefits of data analytics in logistics are undeniable.

As technology continues to advance, the role of data analytics in logistics will only grow in importance. Companies that effectively harness the power of data will be better positioned to navigate the complexities of the modern supply chain, achieve operational excellence, and meet the ever-increasing demands of their customers.

Big Data Applications in Transportation

The integration of big data into transportation systems has revolutionized the way we understand and manage mobility. With the proliferation of sensors, GPS devices, and mobile technologies, an immense volume of data is generated every second. This data, when effectively harnessed, can provide unprecedented insights into traffic patterns, vehicle performance, and commuter behaviors, thereby enhancing the efficiency, safety, and sustainability of transportation networks.

One of the most significant applications of big data in transportation is traffic management. Traditional traffic management systems relied heavily on static models and limited data sources, often leading to suboptimal flow and congestion. Modern systems, however, leverage real-time data from various sources, including GPS devices, traffic cameras, and social media feeds. By analyzing these data streams, algorithms can predict traffic conditions, optimize signal timings, and reroute vehicles to alleviate congestion. For instance, machine learning models can identify patterns in traffic flow, enabling dynamic adjustments to traffic signals to reduce wait times and improve overall traffic efficiency.

Public transportation systems also benefit significantly from big data analytics. Transit authorities can use data from ticketing systems, passenger counters, and mobile applications to monitor ridership patterns and optimize service delivery. Predictive analytics can forecast demand for different routes and times, allowing for better allocation of resources such as buses and trains. Additionally, real-time data can inform passengers of delays, cancellations, or alternative routes, enhancing the user experience and reliability of public transit.

Another critical area is the monitoring and maintenance of transportation infrastructure. Sensors embedded in roads, bridges, and tunnels collect data on structural health and environmental conditions. This information is crucial for predictive maintenance, where potential issues are identified before they become critical, thereby reducing downtime and repair costs. For example, vibration sensors on a bridge can detect anomalies that may indicate structural weaknesses, prompting timely inspections and interventions.

Freight and logistics operations have also been transformed through big data applications. Companies can track shipments in real-time, optimize delivery routes, and manage inventory with greater precision. Data analytics enable the identification of inefficiencies in the supply chain, leading to cost savings and improved delivery times. Advanced algorithms can also forecast demand, allowing for better planning and inventory management. Moreover, data-driven insights can enhance fuel efficiency and reduce carbon emissions by optimizing vehicle loads and routes.

The advent of autonomous vehicles represents another frontier where big data plays a pivotal role. Self-driving cars rely on a multitude of sensors and data inputs to navigate and make decisions in real-time. High-definition maps, traffic data, and environmental conditions are continually processed to ensure safe and efficient operation. Machine learning models are trained on vast datasets to recognize objects, predict movements, and respond to dynamic driving environments. The continuous feedback loop between vehicle sensors and central data repositories enables ongoing improvements in autonomous driving technologies.

Safety is a paramount concern in transportation, and big data analytics contribute significantly to enhancing it. Predictive analytics can identify high-risk areas and times based on historical accident data, enabling targeted interventions, such as increased patrols or improved signage. Real-time monitoring systems can alert authorities to hazardous conditions, such as icy roads or traffic accidents, facilitating rapid response and mitigation.

In sum, the deployment of big data in transportation systems offers transformative benefits across various domains. From traffic management and public transit to infrastructure maintenance and autonomous vehicles, the ability to collect, analyze, and act on vast amounts of data is driving significant advancements. The continuous evolution of data analytics and machine learning technologies promises even greater efficiency and innovation in the future, making transportation systems smarter, safer, and more responsive to the needs of society.

Case Studies of Data-Driven Logistics

The integration of data-driven methodologies into logistics operations has revolutionized the efficiency of supply chain management. This chapter delves into several case studies illustrating the transformative impact of data analytics on logistics, highlighting diverse applications and outcomes.

One prominent example is the deployment of predictive analytics by a global e-commerce giant. By leveraging vast datasets encompassing customer purchasing behavior, seasonal trends, and inventory levels, the company has optimized its inventory management system. Advanced algorithms predict product demand with remarkable accuracy, enabling the company to maintain optimal stock levels. This predictive capability minimizes out-of-stock scenarios and reduces excess inventory, leading to substantial cost savings and improved customer satisfaction.

In another case, a leading logistics service provider has implemented real-time tracking and data analytics to enhance its delivery operations. By equipping its fleet with GPS-enabled devices, the company collects real-time data on vehicle locations, traffic conditions, and delivery times. This data is then analyzed to identify patterns and anomalies, allowing for dynamic route optimization. The result is a significant reduction in delivery times and fuel consumption, which not only cuts operational costs but also lessens the environmental impact of the logistics operations.

A third example involves a major retail chain that has harnessed the power of big data to streamline its supply chain. By integrating data from various sources, including point-of-sale systems, supplier databases, and market research, the company has developed a comprehensive view of its supply chain dynamics. Machine learning models analyze this data to forecast demand, manage supplier relationships, and optimize replenishment schedules. This holistic approach has led to a more resilient supply chain, capable of adapting quickly to market changes and disruptions.

The food and beverage industry also offers a compelling case study of data-driven logistics. A leading beverage manufacturer has adopted data analytics to enhance its distribution network. By analyzing historical sales data, weather patterns, and promotional activities, the company predicts demand at different locations with high precision. This enables the company to allocate resources more effectively, ensuring timely deliveries and reducing waste due to spoilage. The integration of IoT sensors in storage and transportation units' further aids in monitoring the conditions of perishable goods, ensuring quality and compliance with regulatory standards.

An additional case study focuses on the use of blockchain technology in logistics by a multinational shipping company. The company has implemented a blockchain-based system to enhance transparency and traceability in its supply chain. By recording every transaction on an immutable ledger, the company ensures that all parties have access to a single source of truth, reducing disputes and enhancing trust. The use of smart contracts automates various processes, such as customs clearance and payment settlements, thereby speeding up operations and reducing administrative overhead.

These case studies underscore the profound impact of data-driven approaches on logistics. By harnessing the power of data analytics, companies across various industries have achieved significant improvements in efficiency, cost reduction, and customer satisfaction. The adoption of advanced technologies such as predictive analytics, real-time tracking, big data integration, IoT, and blockchain has enabled these companies to transform their logistics operations, setting new benchmarks for the industry. The continuous evolution of data analytics and its applications promises to further revolutionize logistics, offering new opportunities for innovation and growth.

Challenges and Opportunities

The rapid expansion of digital technologies has significantly transformed various sectors, presenting both formidable challenges and remarkable opportunities. One of the primary challenges is the issue of cybersecurity. As more data is transmitted over digital networks, the risk of cyber-attacks increases exponentially. Organizations must invest heavily in advanced security measures to protect sensitive information from breaches. This necessitates continuous updates to security protocols and systems, which can be both costly and resource-intensive.

Another significant challenge is the digital divide. Access to high-speed internet and digital tools is not uniformly distributed, leading to disparities between different regions and socio-economic groups. This gap can exacerbate existing inequalities, as those without access to digital resources are left behind in education, job opportunities, and social services. Bridging this divide requires concerted efforts from governments, private sectors, and non-profit organizations to provide affordable and reliable internet access to underserved communities.

The rise of digital technologies also brings about concerns related to privacy. With the proliferation of data collection practices, individuals' personal information is more vulnerable to misuse. Striking a balance between leveraging data for innovation and respecting privacy rights is a complex task. Regulatory frameworks such as the General Data Protection Regulation (GDPR) in Europe are steps in the right direction, but global consensus and cooperation are necessary to address privacy issues effectively.

Despite these challenges, the digital landscape offers immense opportunities. One of the most significant advantages is the democratization of information. Digital platforms enable the widespread dissemination of knowledge, allowing individuals from diverse backgrounds to access educational resources, scientific research, and cultural content. This has the potential to foster a more informed and educated global populace.

In the realm of business, digital technologies have revolutionized operations and strategies. E-commerce platforms have opened new markets, enabling businesses to reach customers worldwide with minimal overhead costs. Additionally, digital tools such as data analytics and artificial intelligence allow companies to optimize their operations, enhance customer experiences, and make data-driven decisions. This can lead to increased efficiency, productivity, and profitability.

Healthcare is another sector that stands to benefit significantly from digital advancements. Telemedicine, for instance, has made healthcare services more accessible, especially in remote areas. Digital health records streamline patient care and improve coordination among healthcare providers. Moreover, wearable devices and health apps empower individuals to actively monitor their health and well-being actively.

The digital transformation also impacts the labor market. While there are concerns about job displacement due to automation and artificial intelligence, new job opportunities are emerging in tech-related fields. There is a growing demand for skills in software development, data science, cybersecurity, and digital marketing. Upskilling and reskilling initiatives are crucial to prepare the workforce for these new roles and ensure a smooth transition.

Environmental sustainability can also benefit from digital technologies. Smart grids, for example, optimize energy distribution and reduce waste. Digital platforms facilitate the sharing economy, promoting resource efficiency and reducing consumption. Additionally, data analytics can enhance environmental monitoring and management, aiding in the fight against climate change.

In conclusion, the digital highway presents a complex landscape of challenges and opportunities. Addressing the challenges requires a multi-faceted approach involving technological innovation, regulatory measures, and collaborative efforts. Simultaneously, seizing the opportunities necessitates a forward-thinking mindset and a willingness to adapt to the rapidly evolving digital environment. By navigating these dynamics effectively, societies can harness the full potential of digital technologies to drive progress and improve quality of life.

Chapter 10

Cybersecurity in Road Logistics

Cybersecurity Threats in Logistics

The digital transformation of the logistics industry has brought about significant advancements in efficiency, transparency, and customer satisfaction. However, this increased reliance on digital technologies also exposes the sector to a myriad of cybersecurity threats. These threats pose substantial risks to the integrity, availability, and confidentiality of logistics operations, necessitating a comprehensive understanding and proactive approach to cybersecurity.

One of the most prevalent threats in the logistics sector is ransomware. This type of malware encrypts the data of an organization, rendering it inaccessible until a ransom is paid. Logistics companies are particularly vulnerable due to their reliance on real-time data and the critical nature of their operations. A successful ransomware attack can halt operations, leading to significant financial losses and reputational damage. The interconnectedness of logistics networks means that an attack on one company can have cascading effects throughout the supply chain.

Phishing attacks are another significant concern. These attacks typically involve fraudulent emails designed to trick employees into revealing sensitive information or installing malicious software. In the logistics sector, phishing can lead to unauthorized access to critical systems, allowing attackers to steal valuable data or disrupt operations. The human element remains the weakest link in cybersecurity, making employee training and awareness crucial components of a robust cybersecurity strategy.

Supply chain attacks represent a growing threat. These attacks target less secure elements within the supply chain to gain access to more secure systems. For instance, an attacker might compromise a smaller supplier with weaker defenses to infiltrate a larger logistics company. Given the complex and interconnected nature of supply chains, such attacks can be challenging to detect and mitigate. Ensuring that all partners and suppliers adhere to stringent cybersecurity standards is essential for safeguarding the entire supply chain.

IoT devices, increasingly used in logistics for tracking and monitoring, present another vector for cyberattacks. These devices often lack robust security features, making them susceptible to hacking. Compromised IoT devices can be used to launch attacks on larger networks, exfiltrate data, or disrupt operations. Implementing strong security measures for IoT devices, such as regular updates and network segmentation, is essential for protecting logistics operations.

Data breaches are a critical concern, as logistics companies handle vast amounts of sensitive information, including customer data, shipment details, and financial records. Unauthorized access to this data can lead to identity theft, financial fraud, and competitive disadvantage. Protecting data through encryption, access controls, and regular audits is essential for maintaining the confidentiality and integrity of information.

The increasing use of cloud services in logistics also introduces new cybersecurity challenges. While cloud providers offer robust security measures, the responsibility for securing data in the cloud is shared between the provider and the user. Misconfigurations, inadequate access controls, and lack of visibility can lead to data breaches and other security incidents. Adopting best practices for cloud security, such as strong authentication, regular monitoring, and secure configuration, is necessary for protecting cloud-based logistics operations.

The evolving nature of cyber threats requires logistics companies to adopt a proactive and adaptive approach to cybersecurity. This includes investing in advanced security technologies, regular training and awareness programs for employees, and establishing incident response plans. Collaboration with industry partners, cybersecurity experts, and regulatory bodies is also crucial to staying ahead of emerging threats and ensuring the resilience of logistics operations in the digital age.

Technological Solutions for Cybersecurity

The increasing reliance on digital systems necessitates robust cybersecurity measures to protect sensitive information and maintain the integrity of technological infrastructures. Technological solutions for cybersecurity are multifaceted and continuously evolving to counteract sophisticated cyber threats.

Encryption is a cornerstone of cybersecurity, ensuring that data transmitted across networks remains confidential and unaltered. Advanced Encryption Standards (AES) and Public Key Infrastructure (PKI) are widely adopted mechanisms. AES employs symmetric key algorithms, providing a high level of security for data at rest and in transit. PKI, on the other hand, utilizes asymmetric cryptography to authenticate and secure communications, ensuring that only authorized entities can access the encrypted data.

Another critical component is Intrusion Detection and Prevention Systems (IDPS). These systems monitor network traffic for suspicious activities and potential threats. IDPS can be categorized into network-based and host-based systems. Network-based IDPS analyze data packets traveling across the network, while host-based IDPS focus on individual devices. Machine learning algorithms enhance IDPS by identifying patterns and anomalies indicative of cyber threats, thereby enabling proactive threat mitigation.

Firewalls serve as the first line of defense in network security. They act as a barrier between trusted and untrusted networks, filtering incoming and outgoing traffic based on predefined security rules. Next-generation firewalls (NGFW) extend traditional firewall capabilities by incorporating features such as deep packet inspection, intrusion prevention, and application awareness. NGFWs provide a more comprehensive approach to safeguarding networks against a wide range of cyber threats.

Endpoint security is another vital aspect of cybersecurity. With the proliferation of mobile devices and remote work, securing endpoints has become increasingly challenging. Endpoint Detection and Response (EDR) solutions offer real-time monitoring and analysis of endpoint activities, enabling rapid detection and response to potential threats. EDR platforms leverage behavioral analysis and artificial intelligence to identify and mitigate advanced persistent threats (APT) targeting endpoints.

Identity and Access Management (IAM) systems ensure that only authorized users can access sensitive information and critical systems. IAM encompasses various technologies, including Single Sign-On (SSO), Multi-Factor Authentication (MFA), and Role-Based Access Control (RBAC). SSO simplifies user authentication by allowing access to multiple applications with a single set of credentials. MFA adds an additional layer of security by requiring multiple forms of verification, such as passwords and biometric data. RBAC restricts access based on user roles, minimizing the risk of unauthorized access.

Blockchain technology offers a promising solution for enhancing cybersecurity. Its decentralized and immutable nature makes it difficult for cybercriminals to alter or tamper with data. Blockchain can be employed for secure transactions, identity verification, and supply chain management, among other applications. Smart contracts, powered by blockchain, automate and enforce security protocols, reducing the likelihood of human error and malicious interference.

Artificial Intelligence (AI) and Machine Learning (ML) is revolutionizing cybersecurity by enabling more efficient threat detection and response. AI-driven security solutions can process vast amounts of data, identifying patterns and anomalies that may indicate cyber threats. ML algorithms continuously learn from new data, improving their accuracy and effectiveness over time. These technologies enhance the ability to predict, detect, and mitigate cyber threats, providing a more dynamic and adaptive approach to cybersecurity.

Incorporating these technological solutions is essential for building resilient cybersecurity infrastructures. As cyber threats continue to evolve, so too must the technologies and strategies employed to counter them. Ongoing research and development in the field of cybersecurity are critical for staying ahead of cyber adversaries and ensuring the protection of digital assets.

Case Studies of Cybersecurity in Logistics

The advent of digital technologies has revolutionized the logistics sector, enhancing efficiency, transparency, and reliability. However, this digital transformation has also introduced significant cybersecurity challenges. Analyzing real-world case studies provides valuable insights into the nature of these challenges and the strategies employed to mitigate them.

Maersk, the largest shipping company in the world, suffered a loss of around $300 million due to the NotPetya ransomware attack in June 2017. The malware, initially intended to disrupt Ukrainian infrastructure, quickly spread globally and affected Maersk's operations. The attack crippled the company's IT systems, requiring the reinstallation of 4,000 servers, 45,000 PCs, and 2,500 applications. Despite these extensive damages, Maersk's response highlighted the importance of robust incident response plans and the resilience of their business continuity strategies. The company managed to restore operations within ten days, a testament to their preparedness and the effectiveness of their crisis management protocols.

Another significant incident occurred with FedEx's subsidiary, TNT Express, which also suffered from the NotPetya ransomware. TNT Express experienced severe operational disruptions, leading to a temporary halt in services. The attack underscored the vulnerabilities within the logistics sector, particularly the reliance on interconnected IT systems. FedEx's response involved substantial investments in cybersecurity measures and the overhaul of their IT infrastructure to prevent future breaches. This case illustrates the critical need for continuous monitoring and updating of cybersecurity practices to cope with evolving threats.

The case of the Port of San Diego provides an example of how cyberattacks can target critical infrastructure. In September 2018, the port experienced a ransomware attack that disrupted its IT systems and public services. The port authorities responded by collaborating with various cybersecurity agencies, including the FBI and the Department of Homeland Security, to mitigate the impact. This incident highlighted the importance of inter-agency cooperation and the role of public-private partnerships in responding to cyber threats. The port's experience underscores the necessity of having comprehensive cybersecurity frameworks and the ability to mobilize resources quickly in the face of cyber incidents.

In another instance, the logistics company Expeditors International faced a cyberattack in February 2022 that led to the shutdown of their global operations. The attack demonstrated the potential for cyber incidents to cause widespread disruption in supply chains. Expeditors International's response involved activating their cybersecurity incident response plan, which included isolating affected systems, engaging cybersecurity experts, and communicating transparently with stakeholders. The company's recovery efforts emphasized the importance of having a well-prepared incident response plan and the ability to execute it efficiently.

These case studies reveal several key lessons for the logistics sector. Firstly, the importance of having robust cybersecurity measures in place cannot be overstated. This includes not only technical defenses but also comprehensive incident response plans and regular cybersecurity training for employees. Secondly, the interconnected nature of modern logistics operations means that a cyberattack on one part of the supply chain can have far-reaching consequences. Therefore, companies must adopt a holistic approach to cybersecurity, considering the entire supply chain and collaborating with partners to enhance overall security.

Furthermore, these incidents illustrate the evolving nature of cyber threats and the need for continuous adaptation of cybersecurity strategies. Regularly updating systems, conducting vulnerability assessments, and staying informed about the latest threat landscapes are crucial practices for mitigating risks. Finally, the role of public-private partnerships and inter-agency cooperation is vital in responding to and recovering from cyber incidents. Collaborative efforts can enhance the resilience of the logistics sector and ensure a coordinated response to cyber threats.

In conclusion, these case studies highlight the critical importance of cybersecurity in the logistics sector. By learning from past incidents and adopting proactive measures, companies can better protect their operations and ensure the continued efficiency and reliability of global supply chains.

Future Trends in Cybersecurity

The landscape of cybersecurity is continually evolving, driven by advancements in technology and the increasing sophistication of cyber threats. As we look ahead, several key trends are poised to shape the future of cybersecurity, necessitating adaptive strategies and innovative solutions.

One of the most significant trends is the integration of Artificial Intelligence (AI) and Machine Learning (ML) into cybersecurity frameworks. These technologies offer the potential to enhance threat detection and response times by analyzing vast amounts of data more quickly and accurately than traditional methods. AI and ML can identify patterns and anomalies that may indicate a cyber threat, thereby enabling preemptive actions. Furthermore, these technologies can improve the accuracy of threat intelligence, reducing the number of false positives and allowing cybersecurity professionals to focus on genuine threats.

The proliferation of Internet of Things (IoT) devices presents both a critical challenge and an opportunity for cybersecurity. As the number of connected devices continues to grow, so does the attack surface for potential cyber threats. Ensuring the security of IoT devices requires robust encryption methods, secure communication protocols, and regular updates to address vulnerabilities. The development of standardized security frameworks for IoT devices will be essential in mitigating risks associated with this trend.

Quantum computing represents a double-edged sword for cybersecurity. On one hand, quantum computers have the potential to break existing encryption algorithms, posing a significant threat to data security. On the other hand, quantum computing can also be harnessed to develop new cryptographic techniques that are more secure against quantum attacks. The transition to quantum-resistant encryption methods will be a crucial focus for the cybersecurity community in the coming years.

Zero Trust Architecture (ZTA) is gaining traction as a fundamental principle in cybersecurity strategy. Unlike traditional security models that rely on perimeter defenses, ZTA operates on the premise that threats can originate from both outside and inside the network. Therefore, it advocates for continuous verification of user and device identities, regardless of their location. Implementing ZTA involves stringent access controls, micro-segmentation of networks, and continuous monitoring of user activity to detect and respond to threats in real time.

The increasing reliance on cloud computing services necessitates robust cloud security measures. As organizations migrate their data and applications to the cloud, they must ensure that cloud service providers adhere to stringent security standards. This includes data encryption, secure access controls, and regular security assessments. Additionally, organizations must implement their own security measures to protect sensitive data in the cloud, such as using multi-factor authentication and conducting regular security audits.

Human factors remain a critical aspect of cybersecurity. Despite technological advancements, human error continues to be a leading cause of security breaches. Therefore, ongoing education and training for employees are essential to foster a culture of cybersecurity awareness. This includes regular phishing simulations, security awareness programs, and clear communication of security policies and procedures.

Regulatory compliance will continue to play a vital role in shaping cybersecurity practices. Governments and regulatory bodies are increasingly enacting laws and regulations to protect data privacy and security. Organizations must stay abreast of these regulatory changes and ensure compliance to avoid penalties and protect their reputation.

In conclusion, the future of cybersecurity will be characterized by the integration of advanced technologies, the need for standardized security frameworks, and a focus on human factors and regulatory compliance. By staying ahead of these trends, organizations can better protect themselves against the ever-evolving landscape of cyber threats.

Chapter 11

Human Factors and Workforce Transformation

Impact of Technology on Workforce

The advent of digital technology has profoundly transformed the landscape of the modern workforce. This transformation is characterized by the integration of advanced technologies such as Artificial Intelligence (AI), machine learning, automation, and data analytics into various sectors. These technologies have redefined job roles, skill requirements, and organizational structures, leading to both opportunities and challenges for workers and employers alike.

Automation and AI have significantly enhanced productivity and efficiency across industries. Routine and repetitive tasks, once performed by human workers, are increasingly being handled by machines. This shift has resulted in the displacement of certain job categories, particularly those involving manual labor or simple computational tasks. However, it has also created new job opportunities in fields such as robotics, AI programming, and data analysis. The demand for workers with expertise in these areas has surged, necessitating a shift in educational and training paradigms to equip the workforce with the necessary skills.

The digitalization of the workplace has facilitated the rise of remote work and the gig economy. Advances in communication technologies and collaboration tools have enabled employees to work from virtually anywhere, breaking down geographical barriers and providing greater flexibility. This has led to the proliferation of freelance and contract work, allowing individuals to offer their services on a project-by-project basis. While this model offers greater autonomy and work-life balance for some, it also presents challenges related to job security, benefits, and worker protections.

Data analytics and big data have become integral to decision-making processes within organizations. The ability to collect, process, and analyze vast amounts of data in real-time has empowered businesses to make more informed and strategic decisions. This data-driven approach has streamlined operations, optimized resource allocation, and enhanced customer experiences. Consequently, there is a growing need for data scientists, analysts, and professionals skilled in interpreting and leveraging data to drive business outcomes.

The integration of technology in the workforce has also necessitated a cultural shift within organizations. Companies must foster a culture of continuous learning and adaptability to keep pace with technological advancements. This involves investing in employee training and development programs to ensure that the workforce remains competitive and capable of leveraging new technologies. Additionally, organizations must address ethical considerations related to technology use, such as data privacy, algorithmic bias, and the impact of automation on employment.

Despite the numerous benefits, the rapid pace of technological change has also raised concerns about widening skill gaps and inequality. Workers who lack access to digital skills training or are unable to adapt to new technologies risk being left behind. This digital divide underscores the importance of inclusive policies and initiatives aimed at bridging the gap and ensuring that all workers can participate in the digital economy.

The impact of technology on the workforce is multifaceted, encompassing both positive and negative aspects. While technology has the potential to drive economic growth and innovation, it also poses significant challenges that need to be addressed. By understanding these dynamics and proactively adapting to the changing landscape, organizations and workers can harness the benefits of technology while mitigating its adverse effects. The ongoing evolution of the digital highway will continue to shape the future of work, demanding a concerted effort from all stakeholders to navigate its complexities.

Skills and Training for the Digital Age

The accelerating pace of technological innovation has fundamentally transformed the landscape of skills and training. As industries pivot towards digital integration, the demand for a workforce proficient in digital competencies has surged. This shift necessitates a comprehensive understanding of the skills required and the corresponding training methodologies to prepare individuals for the evolving digital economy.

Central to the digital age is proficiency in data literacy, which encompasses the ability to analyze, interpret, and utilize data effectively. Data literacy is not limited to specialists; it has become a foundational skill across various sectors. Employees are expected to engage with data-driven decision-making processes, necessitating a robust understanding of data analytics tools and methodologies. This includes familiarity with software such as Python, R, and SQL, as well as the capability to interpret complex datasets to derive actionable insights.

In parallel, the digital age has underscored the importance of cybersecurity skills. As organizations increasingly rely on digital infrastructure, the risk of cyber threats has escalated. This has created a critical need for expertise in cybersecurity measures, including threat detection, risk management, and the implementation of robust security protocols. Training programs must, therefore incorporate comprehensive modules on cybersecurity principles, ethical hacking, and incident response strategies to equip the workforce with the necessary skills to safeguard digital assets.

Moreover, the advent of Artificial Intelligence (AI) and Machine Learning (ML) has introduced new dimensions to skill requirements. Proficiency in AI and ML is becoming indispensable, with applications spanning numerous industries such as healthcare, finance, and manufacturing. Training in this domain involves understanding algorithms, neural networks, and the development of predictive models. This requires a solid foundation in mathematics and statistics, as well as practical experience with platforms like TensorFlow and PyTorch.

Digital literacy also extends to soft skills, which are critical in a collaborative and dynamic work environment. Communication, problem-solving, and adaptability are paramount as teams navigate complex digital ecosystems. Training programs must therefore integrate modules that foster these soft skills, ensuring that employees can effectively collaborate and innovate within digital frameworks.

The mode of delivery for digital skills training has itself undergone transformation. Traditional classroom-based training is increasingly supplemented or replaced by online learning platforms. These platforms offer flexibility and accessibility, enabling individuals to acquire new skills at their own pace. Massive Open Online Courses (MOOCs), webinars, and virtual workshops have become prevalent, providing a diverse array of learning opportunities. These digital learning environments are often enhanced by interactive elements such as simulations and gamification, which enhance engagement and retention.

Corporate training programs are also evolving to keep pace with digital advancements. Many organizations are adopting continuous learning models, where employees regularly update their skills to stay relevant. This approach is facilitated by Learning Management Systems (LMS) that track progress and provide personalized learning paths. Additionally, partnerships with educational institutions and tech companies are becoming common, fostering a collaborative approach to skill development.

The rapid integration of digital technologies into various sectors underscores the urgency for a well-equipped workforce. Ensuring that individuals possess the requisite digital skills and receive appropriate training is paramount for thriving in the digital age. This necessitates a multi-faceted approach that combines technical proficiency with soft skills, delivered through innovative and accessible training platforms. Only through such comprehensive preparation can the workforce effectively navigate and excel in the digital economy.

Case Studies of Workforce Adaptation

The rapid proliferation of digital technologies has necessitated significant shifts in workforce dynamics across various industries. This section delves into specific instances where organizations have successfully navigated the complexities of digital transformation by fostering workforce adaptation. Through these case studies, we aim to elucidate the strategies and methodologies employed to mitigate resistance and enhance workforce agility.

One notable example is the transformation within a leading global financial services firm. Facing increasing competition from fintech startups, the firm initiated a comprehensive digital overhaul. This involved the integration of advanced analytics, artificial intelligence (AI), and blockchain technologies. The workforce adaptation strategy was underpinned by a robust training program designed to upskill employees in these emerging technologies. The firm established a digital academy offering modular courses tailored to different job functions. Employees were incentivized to participate through a combination of career advancement opportunities and financial rewards. This approach not only equipped the workforce with the necessary technical skills but also fostered a culture of continuous learning and innovation.

In the manufacturing sector, a German automotive company provides another illustrative case. The company adopted Industry 4.0 principles, incorporating Internet of Things (IoT) devices, robotics, and big data analytics into its production processes. The workforce adaptation strategy centered around a collaborative approach involving both management and labor unions. A joint task force was created to oversee the transition, ensuring that workers' concerns were addressed and that the implementation was as seamless as possible. The company also invested in reskilling programs, focusing on technical and digital literacy. By fostering a participatory environment and providing clear pathways for skill development, the company was able to maintain high levels of employee engagement and productivity during the transition.

A different perspective can be observed in the retail industry, where a major multinational corporation faced the challenge of integrating e-commerce with traditional brick-and-mortar operations. The company launched a dual-channel strategy, emphasizing the importance of digital literacy across all levels of the organization. A key component of their workforce adaptation plan was the deployment of digital ambassadors—employees who received specialized training in e-commerce technologies and were then tasked with disseminating this knowledge to their peers. This peer-to-peer training model proved effective in accelerating the adoption of new technologies and minimizing disruption. Additionally, the company implemented a feedback loop mechanism, allowing employees to voice concerns and suggestions, which were then used to refine the adaptation process continuously.

In the healthcare sector, a large hospital network in the United States embarked on a digital transformation journey to improve patient care and operational efficiency. The adoption of electronic health records (EHR), telemedicine, and AI-driven diagnostics necessitated a significant shift in workforce competencies. The hospital network introduced a phased training program, starting with basic digital literacy and gradually progressing to more advanced applications. Importantly, the training was tailored to the specific needs and roles of different healthcare professionals, from administrative staff to clinicians. This targeted approach ensured that each segment of the workforce could effectively integrate digital tools into their daily routines. The hospital network also established a digital support team to provide ongoing assistance and troubleshoot issues, thereby reducing resistance and fostering a smooth transition.

These case studies underscore the importance of a strategic and inclusive approach to workforce adaptation in the digital age. By investing in targeted training programs, fostering collaborative environments, and leveraging internal resources, organizations can successfully navigate the challenges posed by digital transformation. The lessons derived from these examples offer valuable insights for other entities seeking to enhance their workforce's adaptability and resilience in an increasingly digital world.

Future Workforce Trends

The rapid evolution of digital technologies is reshaping the landscape of workforce dynamics, introducing new paradigms that will define future employment. The integration of artificial intelligence (AI), machine learning, and automation is expected to significantly alter job roles and skill requirements. This transformation necessitates a comprehensive understanding of emerging trends to prepare for the future workforce effectively.

One of the most significant trends is the increasing reliance on AI and automation. These technologies are poised to handle routine and repetitive tasks, leading to enhanced efficiency and productivity. However, this shift will also result in the displacement of certain job categories, particularly those involving manual or clerical work. Consequently, there is a pressing need for reskilling and upskilling programs to help workers, transition into roles that require more complex cognitive abilities and problem-solving skills. The focus will be on developing expertise in areas, where human judgment and creativity are irreplaceable.

The advent of remote work, accelerated by the COVID-19 pandemic, is another critical trend shaping the future workforce. Remote work has demonstrated that many tasks can be performed outside traditional office settings, offering flexibility and work-life balance. This shift is likely to persist, with hybrid work models becoming more prevalent. Organizations will need to invest in digital infrastructure and cybersecurity measures to support a distributed workforce effectively. Additionally, management practices will evolve to address the challenges of remote supervision and to foster collaboration and engagement among geographically dispersed teams.

The gig economy is expanding, driven by digital platforms that connect freelancers with short-term job opportunities. This model offers flexibility for workers and provides employers with access to a diverse talent pool. However, it also raises concerns about job security, benefits, and fair compensation. Policymakers and industry leaders must collaborate to create frameworks that protect gig workers' rights while promoting innovation and economic growth.

Data literacy is becoming a critical skill across various industries. As organizations increasingly rely on data-driven decision-making, employees at all levels must be proficient in interpreting and leveraging data. This trend is leading to the emergence of roles such as data scientists, data analysts, and business intelligence experts. Education systems and corporate training programs must adapt to equip individuals with the necessary skills to thrive in a data-centric environment.

The concept of lifelong learning is gaining prominence as technological advancements continually reshape job requirements. Traditional education models are being supplemented by continuous learning initiatives, including online courses, certifications, and professional development programs. Employers are recognizing the value of investing in their workforce's ongoing education to maintain a competitive edge. This shift underscores the importance of fostering a culture of continuous improvement and adaptability.

Ethical considerations are becoming increasingly important as digital technologies permeate the workforce. Issues such as data privacy, algorithmic bias, and the ethical use of AI are at the forefront of discussions. Organizations must establish clear ethical guidelines to navigate these challenges responsibly. Transparency and accountability will be crucial in building trust among employees, customers, and stakeholders.

In summary, the future workforce will be characterized by a dynamic interplay of technology, flexibility, and continuous learning. Preparing for these changes requires a proactive approach to skill development, ethical considerations, and the adoption of innovative work models. Organizations that successfully navigate these trends will be well-positioned to thrive in the digital age.

Chapter 12

Policy and Regulatory Frameworks

Current Policies in Road Logistics

Road logistics has increasingly become a focal point for policymakers, given its critical role in global supply chains and economic activity. Contemporary policies in road logistics are designed to address a myriad of challenges, including environmental sustainability, safety, efficiency, and technological integration. This subchapter explores the current landscape of policies governing road logistics, emphasizing regulatory frameworks, sustainability initiatives, safety standards, and the integration of digital technologies.

Regulatory frameworks are the backbone of road logistics, ensuring that operations are conducted within legal and ethical boundaries. In many regions, these regulations are becoming more stringent to address the growing concerns over environmental impact and road safety. For instance, the European Union's Mobility Package aims to create a fairer and more efficient road transport sector by setting clear rules on driving times, rest periods, and cabotage operations. Similarly, the United States has implemented the Electronic Logging Device (ELD) mandate, which requires commercial motor vehicle drivers to use electronic devices to record driving hours, thereby reducing the risk of fatigue-related accidents.

Sustainability is a paramount concern in current road logistics policies. Governments are increasingly focusing on reducing the carbon footprint of road transport. The European Green Deal, for instance, targets a 90% reduction in transport emissions by 2050, which includes significant changes to road logistics. Policies promoting the adoption of electric and alternative fuel vehicles are gaining traction. Incentives for the use of electric trucks, such as tax breaks and subsidies, are becoming more common. Additionally, low-emission zones in urban areas restrict the entry of high-emission vehicles, compelling logistics companies to upgrade their fleets to more environmentally friendly options.

Safety standards remain a critical aspect of road logistics policies. The introduction of advanced driver-assistance systems (ADAS) and autonomous driving technologies are at the forefront of policy discussions. Regulatory bodies are working on setting standards for these technologies to ensure they enhance safety without introducing new risks. For instance, the National Highway Traffic Safety Administration (NHTSA) in the United States has been actively involved in developing guidelines for autonomous vehicles. Similarly, the European Union has established the General Safety Regulation, which mandates the inclusion of various safety features in new vehicles, such as intelligent speed assistance and advanced emergency braking systems.

The integration of digital technologies is revolutionizing road logistics. Policies are increasingly supportive of digital transformation, recognizing its potential to enhance efficiency and transparency. The adoption of telematics, blockchain, and Internet of Things (IoT) technologies is being encouraged through various initiatives. For example, the European Union's Digital Transport and Logistics Forum (DTLF) aims to facilitate the digitalization of transport and logistics, enhancing data sharing and interoperability among stakeholders. In the United States, the Federal Motor Carrier Safety Administration (FMCSA) supports the deployment of technology to improve logistics operations, such as through the Automated Commercial Environment (ACE) system, which streamlines data submission and processing.

In summary, current policies in road logistics are multifaceted, addressing regulatory compliance, environmental sustainability, safety, and technological advancements. These policies are evolving to meet the demands of a rapidly changing industry, ensuring that road logistics remains efficient, safe, and sustainable. As governments and regulatory bodies continue to develop and implement these policies, the road logistics sector is poised to undergo significant transformations, driven by the dual imperatives of innovation and regulation.

Regulatory Challenges and Opportunities

The rapid advancement of digital technologies and their integration into various sectors have presented both challenges and opportunities from a regulatory perspective. The convergence of digital platforms, artificial intelligence, and the Internet of Things (IoT) has necessitated a reevaluation of existing regulatory frameworks to ensure they remain relevant and effective. This subchapter delves into the complexities of regulating the digital highway, highlighting the intricate balance between fostering innovation and protecting public interests.

One of the primary challenges in regulating digital technologies is the pace at which these innovations evolve. Traditional regulatory processes, often characterized by lengthy deliberations and bureaucratic procedures, struggle to keep up with the rapid development cycles of digital products and services. This lag can lead to outdated regulations that fail to address the nuances of new technologies or, conversely, overly restrictive measures that stifle innovation. Policymakers are thus faced with the arduous task of designing agile regulatory frameworks that can adapt to the fast-changing digital landscape.

Data privacy and security are paramount concerns in the digital age. The proliferation of data-driven technologies has raised significant issues regarding the collection, storage, and usage of personal information. Regulatory bodies must ensure that robust data protection laws are in place to safeguard individuals' privacy rights while enabling the beneficial uses of data. The General Data Protection Regulation (GDPR) in the European Union serves as a notable example of comprehensive data protection legislation. However, the global nature of digital technologies necessitates international cooperation and harmonization of data privacy standards to address cross-border data flows effectively.

The rise of artificial intelligence (AI) introduces another layer of regulatory complexity. AI systems, particularly those employing machine learning algorithms, can exhibit behaviors that are difficult to predict or explain. Ensuring transparency and accountability in AI decision-making processes is critical to maintaining public trust. Regulatory frameworks must establish clear guidelines for the ethical development and deployment of AI, addressing issues such as algorithmic bias, accountability, and the potential for job displacement due to automation.

The digital economy's reliance on platform-based business models also poses unique regulatory challenges. Dominant digital platforms, such as social media giants and e-commerce marketplaces, wield significant market power that can stifle competition and innovation. Antitrust and competition laws need to be revisited and potentially reformed to address the monopolistic tendencies of these platforms. Additionally, the role of content moderation on digital platforms has come under scrutiny, necessitating a careful balance between preventing harmful content and upholding freedom of expression.

Despite these challenges, the digital highway presents numerous regulatory opportunities. Proactive regulatory approaches can stimulate innovation by providing clear guidelines and reducing uncertainty for businesses. Regulatory sandboxes, for example, allow companies to test new technologies in a controlled environment under the supervision of regulators. This fosters a collaborative relationship between innovators and policymakers, facilitating the development of regulations that are both effective and conducive to technological advancement.

Furthermore, the digital highway offers the potential for enhanced regulatory efficiency through the use of technology. Regulatory technology (RegTech) solutions, such as automated compliance systems and blockchain-based audit trails, can streamline regulatory processes and improve oversight. These technological tools enable regulators to monitor compliance in real-time, reducing the administrative burden on both regulators and regulated entities.

In navigating the regulatory challenges and opportunities of the digital highway, it is imperative for policymakers to adopt a forward-looking and flexible approach. By embracing technological advancements and fostering collaboration between stakeholders, regulatory frameworks can be designed to support innovation while safeguarding public interests.

Case Studies of Policy Implementation

The examination of policy implementation within the context of digital infrastructure development reveals a multifaceted landscape shaped by varying governmental approaches, socio-economic conditions, and technological advancements. This subchapter delves into specific case studies, highlighting the successes and challenges faced by different nations as they navigate the complexities of establishing and enhancing their digital highways.

One illustrative case is South Korea's aggressive push towards digitalization in the early 2000s. The South Korean government implemented a series of strategic policies aimed at fostering a robust digital economy. Key among these was the "Cyber Korea 21" initiative, which focused on expanding broadband infrastructure, promoting ICT education, and supporting technological innovation. The government's direct investment in high-speed internet infrastructure, combined with favorable regulatory frameworks, facilitated rapid adoption and widespread usage. By 2005, South Korea had achieved one of the highest broadband penetration rates globally, positioning itself as a leader in digital infrastructure and innovation.

In contrast, the implementation of digital infrastructure policies in India presents a different narrative. **The Digital India program, launched in 2015, aimed to transform India into a digitally empowered society and knowledge economy**. This initiative encompassed a broad spectrum of objectives, including improving internet connectivity, enhancing digital literacy, and delivering government services electronically. However, the implementation faced significant hurdles, such as inadequate infrastructure in rural areas, digital literacy gaps, and regulatory challenges. Despite these obstacles, the program has made notable progress in increasing internet penetration and facilitating digital financial services, demonstrating the potential for large-scale digital transformation in a diverse and populous nation.

Another pertinent example is the European Union's Digital Agenda, part of the Europe 2020 strategy. This policy framework sought to ensure that every European citizen could access high-speed internet by 2020. The implementation across member states varied significantly due to differences in economic resources, existing infrastructure, and political will. Countries like Sweden and the Netherlands, with strong pre-existing digital frameworks and substantial investment capacity, achieved remarkable progress. Conversely, nations with less developed infrastructure and economic constraints faced slower advancements. This disparity underscores the importance of tailored policy approaches that consider the specific needs and conditions of individual regions within a broader strategic framework.

The United States' approach to digital infrastructure policy showcases a more decentralized model, with significant involvement from the private sector. The Federal Communications Commission (FCC) has played a pivotal role in setting regulatory standards and incentivizing broadband deployment through initiatives such as the Connect America Fund. This program aimed to expand internet access in underserved rural areas by providing financial support to telecommunications companies. While this market-driven approach has fostered innovation and competition, it has also led to uneven access and quality of service across different regions, highlighting the challenges of balancing market dynamics with equitable access goals.

China's digital infrastructure policy is characterized by strong state intervention and strategic planning. The Chinese government's "Internet Plus" strategy, launched in 2015, aimed to integrate internet technologies with traditional industries to foster economic growth and innovation. Government-led investments in fiber optic networks, 5G deployment, and smart city projects have propelled China to the forefront of global digital infrastructure. However, this centralized approach also raises concerns about state control, data privacy, and international competitiveness.

These case studies reflect the diverse strategies and outcomes associated with digital infrastructure policy implementation. They underscore the critical role of government intervention, the importance of adaptable and context-specific approaches, and the interplay between public and private sectors in achieving digitalization goals. The lessons gleaned from these examples provide valuable insights for policymakers aiming to navigate the complex and dynamic landscape of digital infrastructure development.

The Role of Indian Government and Regulatory Bodies in Supporting the Digital Highway

Last year, despite the sweltering Indian summer, my excitement remained high, as I entered the grand conference hall. The government-led conference on the future of road logistics in India was a significant event, bringing together industry leaders and policymakers. I felt privileged to be a part of it.

During the event, Shri Nitin Gadkari, the Hon'ble Minister of Road Transport and Highways, virtually unveiled the **'Digital Twin Strategy for Indian Infrastructure'** report at the GeoSmart Infrastructure 2023 event. As I took my seat among the attendees, I was struck by the impressive array of delegates—representatives from major logistics companies, technology providers, and government officials—gathered for this pivotal discussion.

In his keynote, Mr. Gadkari emphasized, "Indian infrastructure, particularly transport infrastructure, serves as a vital bridge towards transforming India into a developed nation. Embracing modern technologies and harnessing the combined power of geospatial data and digital twins is key to realizing our vision of an inclusive, sustainable, and modern India."

He went on to stress the importance of logistics in achieving the Prime Minister's vision of a USD 5 trillion economy. "To boost exports and reduce imports, we must lower logistics costs from the current 14%-16% to 9% by the end of 2024. This will be made possible through the development of good roads, the use of alternative fuels, and increased reliance on waterways," Mr. Gadkari explained.

Shifting focus to road development, Mr. Gadkari highlighted the strides being made across India, describing them as the foundation of the country's infrastructure modernization. "Our government is committed to building world-class road infrastructure," he stated. "We are constructing over 22 green expressways, which will drastically reduce travel time and fuel consumption, thereby lowering logistics costs. The Delhi-Mumbai Expressway, which is nearing completion, will cut travel time between these two major economic hubs to just 12 hours. This expressway is one of many projects designed to enhance connectivity, improve trade routes, and support the growth of our logistics sector."

Mr. Gadkari further noted, "Efficiency and sustainability go hand in hand. By adopting the Digital Highway, we can improve operations while protecting our planet. The new roads we are building are designed to support electric vehicles and other green technologies, which will help reduce our carbon footprint."

In the days that followed, I delved deeper into the government's plans and engaged with fellow industry professionals, eager to learn from their experiences. Together, we aimed to work towards a brighter, more sustainable future for road logistics in India.

As I walked along the busy streets, reflecting on the conference, Mr. Gadkari's words continued to resonate with me. The vision of the Digital Highway and its potential to transform India's supply chain landscape was clear. With government support and the ULIP Platform, logistics companies could streamline their processes, reduce costs, and minimize environmental impact—paving the way for a more efficient and sustainable future.

Future Policy Directions

As the digital landscape continues to evolve at an unprecedented pace, the formulation of future policy directions becomes pivotal in navigating the complexities and opportunities presented by the digital highway.

Policymakers must address a myriad of challenges including data privacy, cybersecurity, digital equity, and the ethical implications of emerging technologies.

One of the foremost considerations is the protection of data privacy. With the proliferation of digital services, vast amounts of personal data are being collected, stored, and processed. Policies must ensure robust data protection frameworks that safeguard individuals' privacy rights while fostering innovation. This involves updating existing regulations to address new data practices and ensuring transparency in data usage. Implementing stringent data protection standards, such as encryption and anonymization, can mitigate risks associated with data breaches and misuse.

Cybersecurity remains a critical concern as cyber threats continue to escalate in sophistication and frequency. Effective policies must encompass comprehensive cybersecurity strategies that include regular risk assessments, incident response protocols, and international collaboration. Encouraging the adoption of best practices in cybersecurity, such as multi-factor authentication and regular software updates, can enhance the resilience of digital infrastructures. Additionally, fostering public-private partnerships can facilitate the sharing of threat intelligence and the development of innovative security solutions.

Digital equity is another pressing issue that requires immediate attention. The digital divide, characterized by disparities in access to digital technologies and the internet, exacerbates social and economic inequalities. Policies should aim to bridge this divide by promoting affordable and universal access to high-speed internet and digital devices. Investment in digital literacy programs is essential to equip individuals with the skills needed to participate fully in the digital economy. Furthermore, inclusive policies that consider the needs of marginalized communities can ensure that the benefits of digital advancements are equitably distributed.

The ethical implications of emerging technologies, such as artificial intelligence (AI) and machine learning, necessitate careful consideration. Policies must address issues related to algorithmic bias, accountability, and transparency. Establishing ethical guidelines for the development and deployment of AI can prevent discriminatory practices and ensure that these technologies are used responsibly. Encouraging interdisciplinary research and stakeholder engagement can provide diverse perspectives and inform the creation of balanced policies that uphold ethical standards.

International cooperation is crucial in addressing the global nature of digital challenges. Harmonizing regulatory frameworks across jurisdictions can facilitate cross-border data flows and enhance cybersecurity efforts. Collaborative initiatives, such as international treaties and agreements, can promote the sharing of best practices and foster a collective approach to digital policy-making. Additionally, supporting the development of global standards for emerging technologies can ensure interoperability and facilitate innovation on a global scale.

Investment in research and development is essential to keep pace with the rapid advancements in digital technologies. Policies that incentivize innovation, through tax credits and grants, can spur the development of cutting-edge solutions. Public funding for research institutions and collaboration with the private sector can drive progress in key areas such as quantum computing, blockchain, and next-generation networks. Moreover, fostering a culture of continuous learning and adaptation can enable societies to stay ahead of technological trends and effectively address emerging challenges.

In crafting future policy directions for the digital highway, a holistic approach that balances innovation with regulation is imperative. By addressing data privacy, cybersecurity, digital equity, and ethical considerations, policymakers can create an environment that promotes sustainable digital growth and ensures that the benefits of technological advancements are realized by all.

Chapter 13

Global Perspectives on Digital Road Logistics

Digital Logistics in North America

The digital transformation of logistics in North America has redefined the operational landscape, leveraging advanced technologies to enhance efficiency, accuracy, and scalability. The integration of digital tools in logistics has facilitated real-time tracking, predictive analytics, and automated processes, significantly reducing costs and improving service delivery.

One of the primary drivers of digital logistics in North America is the advent of the Internet of Things (IoT). IoT devices, embedded with sensors and connected via the internet, enable the continuous monitoring of goods during transit. This real-time visibility allows companies to track shipments accurately, optimize routes, and respond promptly to potential disruptions. The implementation of IoT in logistics has not only enhanced the transparency of supply chains but also improved inventory management and asset utilization.

Artificial Intelligence (AI) and Machine Learning (ML) are also pivotal in transforming logistics operations. AI algorithms analyze vast amounts of data to provide insights into demand forecasting, route optimization, and predictive maintenance. ML models learn from historical data to predict future trends, enabling logistics companies to anticipate market fluctuations and adjust their strategies accordingly. These technologies have been instrumental in reducing delivery times, minimizing fuel consumption, and enhancing customer satisfaction.

Blockchain technology is another significant innovation in the digital logistics landscape. By providing a decentralized and immutable ledger, blockchain ensures the security and integrity of transactional data. This technology is particularly valuable in enhancing the traceability of goods, verifying the authenticity of products, and reducing the risk of fraud. In North America, blockchain has been increasingly adopted to streamline customs procedures, improve supply chain transparency, and foster trust among stakeholders.

The integration of autonomous vehicles and drones represents a forward-looking aspect of digital logistics. Autonomous trucks and delivery drones are being tested and deployed to address the growing demand for faster and more reliable delivery services. These technologies promise to reduce labor costs, enhance delivery speed, and mitigate the impact of driver shortages. In North America, several pilot projects are underway to assess the feasibility and scalability of autonomous logistics solutions.

Cloud computing has also played a crucial role in the digitalization of logistics. Cloud-based platforms provide scalable and flexible infrastructure for managing logistics operations. These platforms enable seamless integration of various logistics functions, from inventory management to order processing and customer service. The adoption of cloud computing has facilitated real-time data sharing and collaboration among stakeholders, enhancing the overall efficiency and responsiveness of logistics networks.

The rise of e-commerce has further accelerated the adoption of digital logistics solutions. The surge in online shopping has created a need for efficient and reliable last-mile delivery services. Digital technologies such as route optimization software, automated sorting systems, and real-time tracking apps have been deployed to meet this demand. In North America, logistics companies are increasingly investing in digital infrastructure to support the growing e-commerce market and ensure timely delivery of goods.

The shift towards digital logistics in North America is supported by robust regulatory frameworks and industry standards. Government initiatives and industry collaborations aim to promote the adoption of digital technologies and ensure the interoperability of systems. These efforts are crucial in addressing the challenges of data security, privacy, and standardization in the digital logistics ecosystem.

In conclusion, the digitalization of logistics in North America is driving significant improvements in operational efficiency and customer satisfaction. The integration of IoT, AI, blockchain, autonomous vehicles, and cloud computing is reshaping the logistics landscape, offering new opportunities for innovation and growth. As the digital transformation continues, logistics companies must adapt to the evolving technological landscape to remain competitive and responsive to market demands.

Digital Logistics in Europe

The integration of digital technologies within the logistics sector in Europe has undergone significant transformations over the past decade. The advent of advanced technologies such as the Internet of Things (IoT), blockchain, Artificial Intelligence (AI), and big data analytics has revolutionized the way goods are transported, stored, and managed across the continent. This subchapter delves into the key components and impacts of digital logistics in Europe, highlighting the technological advancements and their implications for efficiency, transparency, and sustainability.

IoT has played a pivotal role in enhancing the operational efficiency of logistics in Europe. The deployment of IoT-enabled sensors and devices allows for real-time tracking and monitoring of goods throughout the supply chain. This real-time visibility aids in reducing delays, optimizing routes, and ensuring the timely delivery of products. For instance, temperature-sensitive goods such as pharmaceuticals and perishable food items can be monitored continuously to maintain their integrity, thereby reducing spoilage and ensuring compliance with regulatory standards.

Blockchain technology has emerged as a game-changer in providing transparency and security in logistics operations. By creating immutable and decentralized ledgers, blockchain ensures that all transactions and movements of goods are recorded accurately and transparently. This transparency is crucial in combating fraud, reducing paperwork, and enhancing trust among stakeholders. European logistics companies are increasingly adopting blockchain to streamline customs processes, track the provenance of goods, and ensure the authenticity of products, particularly in industries like luxury goods and pharmaceuticals.

AI and machine learning algorithms are being leveraged to optimize various aspects of logistics, from demand forecasting to route planning. Predictive analytics powered by AI can forecast demand with greater accuracy, enabling companies to manage inventory more efficiently and reduce stockouts or overstock situations. AI-driven route planning tools analyze vast amounts of data, including traffic patterns, weather conditions, and historical delivery times, to determine the most efficient routes for transportation. This not only reduces fuel consumption and emissions but also enhances delivery speed and reliability.

Big data analytics is another critical component in the digital transformation of logistics in Europe. The ability to analyze large datasets allows logistics companies to gain valuable insights into their operations and make data-driven decisions. For example, data analytics can identify bottlenecks in the supply chain, predict maintenance needs for transportation vehicles, and optimize warehouse operations. The insights derived from big data enable companies to enhance their overall efficiency, reduce costs, and improve customer satisfaction.

The digitalization of logistics in Europe also aligns with the continent's sustainability goals. The European Union has set ambitious targets to reduce greenhouse gas emissions, and digital technologies play a crucial role in achieving these targets. IoT and AI can optimize routes and reduce fuel consumption, while blockchain can enhance the traceability of goods, ensuring that they are sourced and transported sustainably. Moreover, digital platforms facilitate the sharing of resources, such as vehicles and warehouses, promoting a circular economy and reducing waste.

The adoption of digital logistics in Europe is not without its challenges. The integration of new technologies requires significant investment and a skilled workforce capable of managing and maintaining these systems. Cybersecurity is another critical concern, as the increasing reliance on digital platforms makes logistics companies vulnerable to cyber-attacks. Regulatory compliance and data privacy issues also need to be addressed to ensure the smooth functioning of digital logistics.

Despite these challenges, the benefits of digital logistics in Europe are undeniable. The ongoing advancements in technology and their integration into the logistics sector promise to enhance efficiency, transparency, and sustainability. European logistics companies that successfully navigate these challenges and embrace digitalization are well-positioned to thrive in an increasingly competitive and dynamic market.

Digital Logistics in Asia

The rapid advancement of digital technologies has significantly transformed logistics operations across Asia. This transformation is driven by the integration of Internet of Things (IoT), artificial intelligence (AI), blockchain, and big data analytics into supply chain management. These technologies have enhanced efficiency, transparency, and reliability, providing a competitive edge to companies operating in the region.

IoT has revolutionized logistics by enabling real-time tracking of goods. Sensors and devices embedded in shipping containers, vehicles, and warehouses collect and transmit data, allowing for continuous monitoring of the location and condition of products. This data-driven approach facilitates proactive decision-making, such as rerouting shipments to avoid delays or adjusting storage conditions to maintain product quality. The implementation of IoT in logistics has reduced operational costs and minimized the risk of lost or damaged goods.

AI plays a crucial role in optimizing logistics processes. Machine learning algorithms analyze vast amounts of data to predict demand, optimize routes, and manage inventory levels. Predictive analytics helps companies anticipate market trends and adjust their logistics strategies accordingly. AI-powered automation, such as autonomous vehicles and robotic process automation (RPA), further streamlines operations by reducing human error and increasing throughput. These advancements have led to significant improvements in delivery speed and accuracy.

Blockchain technology enhances transparency and security in logistics operations. Distributed ledger systems provide an immutable record of transactions, ensuring that all stakeholders have access to a single source of truth. This transparency reduces the risk of fraud and errors, as every transaction is recorded and can be audited. Smart contracts, which execute predefined conditions automatically, facilitate seamless and secure exchanges between parties. Blockchain's potential to enhance trust and efficiency in supply chain management is particularly valuable in complex logistics networks across Asia.

Big data analytics drives informed decision-making in logistics. The vast amount of data generated by digital logistics systems is analyzed to uncover patterns and insights that can improve operational efficiency. For instance, data analytics can reveal bottlenecks in the supply chain, enabling companies to address issues promptly. Additionally, customer data analysis helps logistics providers tailor their services to meet specific needs, enhancing customer satisfaction and loyalty.

The digital transformation of logistics in Asia is supported by significant investments in infrastructure and technology. Governments and private enterprises are collaborating to develop smart logistics hubs, digital trade platforms, and advanced transportation networks. These initiatives aim to create a cohesive and efficient logistics ecosystem that leverages digital technologies to drive economic growth.

Despite the benefits, the adoption of digital logistics in Asia faces challenges. Data privacy and security concerns, regulatory compliance, and the need for skilled professionals are critical issues that must be addressed. Furthermore, the digital divide between urban and rural areas poses a challenge to the widespread implementation of advanced logistics technologies.

To mitigate these challenges, stakeholders must prioritize investments in cybersecurity, regulatory frameworks, and education and training programs. Collaboration between governments, industry players, and academia is essential to create a conducive environment for digital logistics innovation.

The digitalization of logistics in Asia is reshaping the industry's landscape, offering unprecedented opportunities for growth and efficiency. As digital technologies continue to evolve, their integration into logistics operations will further enhance the region's competitiveness in the global market. The ongoing efforts to address challenges and foster innovation will determine the future trajectory of digital logistics in Asia.

Comparative Analysis and Insights

The advent of digital technologies has significantly transformed various sectors, leading to a paradigm shift in how businesses, governments, and individuals operate. This section delves into a comparative analysis of the impact of digital technologies across different industries, drawing insights from empirical data and case studies to understand the multifaceted implications of this transformation.

In the healthcare sector, digital technologies have revolutionized patient care, diagnostics, and treatment methodologies. Telemedicine, for instance, has bridged the gap between patients and healthcare providers, particularly in remote areas. A study by the American Telemedicine Association highlights a 35% increase in patient engagement and a 20% reduction in hospital readmission rates due to telehealth services. Moreover, the integration of Artificial Intelligence (AI) in diagnostic procedures has enhanced accuracy and efficiency. Research by Stanford University indicates that AI algorithms can diagnose certain conditions, such as skin cancer, with an accuracy rate comparable to that of experienced dermatologists.

The financial industry has also undergone a significant transformation, driven by fintech innovations. Blockchain technology, for example, has introduced unprecedented levels of transparency and security in financial transactions. A report by PwC indicates that blockchain can reduce infrastructure costs for financial institutions by up to 30%. Additionally, the rise of digital payment systems has facilitated seamless transactions and financial inclusion. The World Bank reports that digital payments have enabled over 1.2 billion people globally to gain access to financial services, thereby reducing the unbanked population.

In the education sector, digital technologies have democratized access to knowledge and learning resources. The proliferation of online courses and educational platforms has made quality education accessible to a broader demographic. According to a study by the Massachusetts Institute of Technology (MIT), online learning can be as effective as traditional classroom instruction, with students demonstrating comparable performance in assessments. Furthermore, the use of data analytics in education has enabled personalized learning experiences, catering to individual student needs and learning paces.

The retail industry has experienced a seismic shift with the rise of e-commerce. Digital platforms have redefined consumer behavior and shopping patterns. Data from the U.S. Census Bureau shows that e-commerce sales accounted for 14.3% of total retail sales in 2021, reflecting a steady increase from previous years. The integration of AI and machine learning in retail has also enhanced customer experience through personalized recommendations and predictive analytics. A report by McKinsey & Company reveals that personalization can drive a 10-15% increase in revenue for retailers.

Public administration and governance have not been immune to the influence of digital technologies. E-governance initiatives have streamlined government services, making them more accessible and efficient. **The United Nations E-Government Survey 2020 indicates that countries with robust e-governance frameworks have higher citizen satisfaction rates and improved service delivery. Digital identity systems, such as India's Aadhaar, have facilitated social welfare distribution and reduced fraud.**

The comparative analysis across these sectors underscores the transformative potential of digital technologies. However, it also highlights challenges such as data privacy, cybersecurity threats, and the digital divide. Addressing these challenges requires a collaborative approach involving policymakers, industry stakeholders, and the academic community. The insights gleaned from this analysis provide a roadmap for leveraging digital technologies to drive sustainable growth and innovation across various sectors.

Chapter 14

The Road Ahead – Embracing the Future of Road Logistics

As we venture deeper into the digital age, the logistics industry, particularly road logistics, is at the cusp of a significant transformation. "The Digital Highway" is not just a metaphor for the evolving landscape of technology but a reality that road logistics must traverse to remain competitive and sustainable. This chapter delves into the critical trends, emerging technologies, and strategies that will define the future of road logistics as it aligns with the broader digital revolution.

Understanding the Trajectory of Digital Evolution in Road Logistics

Digital transformation in road logistics is intricately linked to the broader advancements in digital technology. Historically, technological growth has followed predictable patterns, such as Moore's Law, which has driven exponential increases in computing power. These advancements have laid the groundwork for the integration of sophisticated digital systems into logistics operations. In road logistics, this has manifested in the widespread adoption of real-time data analytics, predictive modelling, and automated systems that optimize routes, manage fleets, and enhance delivery efficiency.

As road logistics companies digitize their operations, they can anticipate and mitigate potential disruptions, such as traffic delays or environmental hazards. This predictive capability is a direct result of the digital tools that process vast amounts of data, offering insights that were previously unattainable.

The Future of Road Logistics: Key Emerging Technologies

A host of new technologies are poised to improve the efficiency, safety and sustainability of road logistics — not just now but in decades ahead. One of the biggest strides they have taken is Artificial Intelligence (AI) inclusion in logistics processes. AI powered systems have changed the way routes are optimized, maintenance is predicted and fleet is managed Using the intelligence from machine learning models which have been trained upon hundreds of data, AI can determine and predict traffic patterns, give optimized routes on deliveries or even tell when a potential machinery failure might occur therefore reducing downtime hence increasing productivity. Additionally, AI can also use actionable insights to inform logistics managers when making decisions.

One of the most important technologies that will shape many aspects in road logistics is the Internet of Things (IoT). Vehicles and cargo are a rich source of data, often coming directly from IoT devices (sensors, smart tags etc) which communicate in real-time. It allows for observing vehicle functionality, cargo conditions and tracking route status therefore offering unheard of insight into the operations of logistics. IoT sensors, for example, can track the temperature or humidity levels of perishable goods throughout a journey and provide feedback about whether they were shipped within an ideal range. Similarly, IoT can improve transparency on security by monitoring the whereabouts of goods and providing notification to stakeholders in whole part of unauthorized access or deviation from planned route.

The role that autonomous vehicles are poised to play in road logistics is also transformative. Autonomous trucks and delivery vehicles would take human drivers out of the mix, decrease operating costs, and maximize efficiency in delivery networks. Self-driving cars can run 24/7 and haven't need to stop for rest like tired drivers on the fast way so that delivery will be faster. Moreover, AI and machine learning integration on these vehicles provides their capability to traverse through difficult roads along with all feature driven obstacles avoidance automatically with the optimization of fuel consumption that makes a more efficient & sustainable logistics operation.

In road logistics, Blockchain technology is on the rise to improve transparency and security. The Blockchain creates a decentralized ledger which is virtually impossible to change, and thereby gives the ability to record everything that happens in every point of the production chain. Times when This Tech May Be Useful For verifying the integrity of goods — enabling cross-verification with manufacturers to check for authenticity and tracking back from destination all way up to source. The administrative process can also be simplified through Blockchain, which in turn saves them from data handling paperwork and then that will eventually reduce the possibility of human mistakes or fraud.

5G networks will play an essential role in enabling these technologies, which are needed for all of this to work properly. 5G will further enable fast, low-latency communication: key for real-time data sharing to support on-device learning in IoT devices and autonomous vehicles. Greater connectivity will allow logistics companies to manage fleets more efficiently, improve traffic management and deliver faster reactions to changing conditions on roads.

VP: Finally, due to all of the above reasons and road logistics striving hard for sustainability we cannot see any other way than sustainable technologies increasing their importance in this branch. Hybrids and electric vehicles (EVs) are well established in this sector, along with other alternative fuels that enable reduced emissions per mile driven; for logistics operations renewable energy sources continue to be integrated. In addition, battery developments are expanding the range and efficiency of electric vehicles (EVs), making them more feasible for long-distance haulage.

In short, the combined power of AI, IoT autonomous vehicles and increasingly sustainability on Blockchain will usher in a new era for road logistics. These technologies are not only driving operational efficiencies and cost savings, but they bring with them the strength to solve existential problems like global sustainability or supply chain transparency. As the technologies and innovations continue to mature, they are poised to reshape road logistics into a more connected, smarter, and sustainable industry.

How Society Changes Lead to the Effects on Road Logistics

The changes in society are important drivers of transforming the road logistics industry, as consumer behaviour is changing, environmental consciousness grows and new rules create space for more innovative and sustainable solutions. These trends are especially apparent in markets such as India, where urbanization is occurring at an unprecedented rate and the rise of a consumer class for e-commerce giants creates new complexities within logistics ecosystems.

E-commerce is one of the major societal change influencing road logistics. The e-commerce market in India has been exploding on the back of rising internet penetration and an increasing number of people using smartphones. This increase in ecommerce orders means that companies can no longer rely on 7-10 day delivery timelines, but must be able to deliver same or next-day shipping times. For e.g., companies like Flipkart and Amazon who are expanding their logistics networks at breakneck speeds, investing in advanced technologies such as AI for optimal routing or IoT real time tracking to make sure delivers always reach on-time despite India's diverse colonial impediment.

One of the most prominent collective changes is growing awareness and emphasis on sustainability. With more focus placed on environmental awareness, consumers and businesses are pushing for environmentally friendly logistics practices. This, consequently, has created numerous projects in India targeting the reduction of carbon emissions from transport. Take for instance, the increasing adoption of electric vehicles (EVs) by logistics and e-commerce companies in India to cater low mileage last-mile deliveries within urban cities that have higher pollution levels. Also, companies increasingly focus on circular economy models that encourage feedstock recycling and reuse of materials in the supply chain to help cut down waste. And the all these initiatives are in line with international sustainability targets and appeal to a growing demographic of consumers looking for green brands.

The other thing is mega cities and urbanization that are simply changing the face of road logistics. A very common problem in cities like Mumbai, Delhi or Bangalore where the population is too high is of traffic and congestion. On the one hand, urbanisation and depopulation resulted in vehicle-volume congestion not only boosting traffic jams but also stretching delivery times. Logistics companies in India are looking at new ways to overcome these challenges, including replacing the last-mile delivery of bikes with electric scooters (in congested areas) and establishing micro-fulfilment centres near their customers for shorter routes. Cities in India too are exploring Smart City initiatives, and cities such as Pune and Kochi have taken the lead with projects that focus on smart infrastructure to aid traffic management better manage congestion leading thereby contributing towards smoother logistics operations.

Also, societal changes in regulatory form are shaping the road logistics industry. Goods and Services Tax (GST) implementation in India — A Game changer that completely redefined the country]]s legacy fragmented tax system, influencing on how logistics operate. GST implementation meant movement of goods can be easier, quicker also leave larger warehouses with the introduction and ensure businesses need not stock in all states to avoid interstate taxes. This reform has ended the existing logistics operation, reducing transportation cost and improving goods movement across all over country.

Along the way, the COVID-19 pandemic has given us a faster insight into changes in societal trends regarding road logistics. A pandemic that spiked demand for contact-less deliveries and waxed interest on hygiene, safety in the logistics process. Logistics facilities in India swiftly acclimated by integrating contactless delivery protocols and digital payment systems while enforcing personal protective equipment (PPE) among their fleet of drivers. CoViD-19 also stressed the need for supply chains to be resilient and led companies to implement digital technologies such as GPS tracking solutions with intelligent analytics & insights, in order improve visibility of their logistics operations.

To put it all together, changes in the industry road logistics are being experienced due to various societal shifts. Especially for countries like India where consumer behaviours change rapidly as urbanization and regulatory reforms continue appearing again dynamics asserts a need for more effective; sustainable, and resilient solutions from logistics perspective To be agile and effective in a dynamic environment such as the one experienced with this crisis, road logistics companies need to adapt by incorporating new technologies into their operations strategy while adjusting to societal expectations.

How The Trucking Industry Is Getting Ready For Future Trials on The Digital Highway In Road Logistics

As the trucking industry is anchored to road logistics, some are calling this a watershed moment in which digitization in all aspects of our lives will affect everything including jobs and employment, as well as business opportunities. This evolution is being propelled by new technologies, social changes and regulatory measures that necessitate a well thought-out game plan. Nevertheless, the degree to which this transformation is already prepared and moving ahead differs greatly between India and trucking companies in developed countries exemplifying diverse challenges as well as opportunities across international geographies.

The trucking sector in developed countries such as the United States, Germany and Japan is increasingly leveraging Autonomous Vehicles (AV) and other cutting-edge technology. The same can likely be said about the businesses in these regions, which enjoy robust infrastructure and large-scale investments made into R&D while local regulatory environments facilitate AV adoption. The U.S. for instance saw notable availability in autonomous truck testing and deployment, with states such as Arizona and Nevada playing an important role as test bed locations. More importantly, they have the money and tech chops to start testing (and deploying) AVs at scale—and vault themselves ahead on this massive technological curve.

Unlike the other countries, India also has challenges in adoption of autonomous vehicles and rest of the emerging technologies. This makes deploying AVs in the country more challenging due to its varied and sometimes harsh road conditions, as well a less sophisticated infrastructure. The technology to develop AVs is expensive and requires massive infrastructure upgrading, which leads to a high cost of implementation. However, the country has taken some measures to breach them as companies are now embracing telematics and IoT integration leading fleet management thereby operational efficiencies. Unlike other places, in India the focus is more on incremental improvements that use these digital tools to solve imminent problems – like route optimization and fuel efficiency.

The digitalization process and data management are also areas where the developed countries significantly differ from India. The use of advanced telematics systems and data analytics platforms is more widespread than in many countries, such as Germany or the U.S., where truck incumbent companies partly already process high volumes of real-time data. This will allow better route planning, Predictive maintenance and also improved safety protocols. Conversely, Indian trucking companies are starting to deploy these technologies but the integration and data utilization are in nascent stage.

Budget constraints as well limited technical knowledge have been a serious challenge for many small and medium enterprises (SMEs) in India that want to adopt tools build on latest in the digital realm. However, digitalization is starting to take on a life of its own with a new generation Logistics firms rushing to get ahead and push the rest of the industry forward.

In developed countries, for instance, driver shortage is another crucial problem from which India takes a different stance. To ensure a steady influx of truck drivers to replace an aging workforce and as younger generations shunned driving careers for those with more creature comforts, particularly in the U.S. and Europe. This has resulted in major initiatives to automate driving tasks and enhance workplace through technology. Though driver shortage is one of the top concern in India, it only adds up to their woes as drivers poor working conditions/ low wages and no social security. This calls for massive interventions not just with technology but institutional check and balances that would motivate a prospective civil engineer. For example, training programs to develop digital literacy and proper handling of new technologies can help close this gap in the profession making it attractive for younger employees.

While sustainability and environmental policies are becoming stringent worldwide, the methodology being adopted by India is in contrast to that of developed countries. Strict EU emissions laws are driving swift growth in electric trucks and alternate fuels, with governments offering generous grants to help firms transition away from carbon emitting systems. As an example, the European Green Deal imposes strict requirements for a reduction in greenhouse gas emissions thereby spurring rapid electrification of logistics fleets. Although environmentally conscious is on the rise in India, and there are a few electric vehicles operating plus other alternatives for alternative fuel models it's still at nascent stage. Although the Indian government has introduced policies like Faster Adoption and Manufacturing of Hybrid and Electric Vehicles (FAME) in India to promote EVs, it is building up slowly because infrastructure hitches like lack of charging stations keep electric trucks from taking over massively due to high costs. Nevertheless, a few Indian logistics companies have started to embrace EVs for last-mile deliveries — mainly in urban areas that experience high levels of pollution.

The regulatory environments are at stark variance in India as compared to most developed countries. The regulatory framework in the EU and also U.S. are comprehensive, very clear regarding emissions standards safety or data protection as well again highly sophisticated compared to other regions of world This form of regulations creates a framework within which companies can be more comfortable innovating and adopting new technologies. In India, the regulations are changing on a continuous basis and hence reforms such as implementation of Electronic Logging Devices (ELDs) in trucks or GST forces us to take them into consideration for future reference. But the regulatory landscape is still fragmented, with different standards in different states making it difficult for truckers running nationwide to comply. While India is in the process of overhauling its regulatory framework, it has an opportunity to put together a more uniform environment for drone-based logistics.

One of the areas in which experiences between India and developed countries diverge is that of supply chain resilience. The COVID-19 pandemic exposed weaknesses in supply chains across the world, with trucking companies finding themselves at a crossroads to adopt more resilient measures. There has been an emphasis in developed countries on strengthening the digital infrastructure, including real-time sensing and analytics for increased supply chain transparency and responsiveness. India, on the other hand has been working to implement digital tools as best they can but have dealing with greater issues — infrastructure and drivers availability when there is a disruption. How well the Indian trucking industry is prepared to deal with these challenges will determine how successfully India can keep goods moving across its roads.

The trucking industry is facing the same challenges on digital front in both India and developed countries, but how these problems are looked at completely different. On the one hand, developed nations with more comprehensive infrastructure and regulatory support can leapfrog to advanced technologies in a relatively shorter time-frame propelling themselves as digital-first countries of this new-age. On the other hand, India is sailing through a nuanced path where technological adoption has to be balanced with infrastructure creation and regulatory modernization. But, as in so many other areas of the country forward thinking on digitalization, sustainability and workforce development could see India's trucking industry taking significant steps forwards. Success will depend on the extent to which these new technologies and societal changes may be harnessed to realise a more resilient, efficient and sustainable road logistics sector that can fully capitalise upon 'The Digital Highway'.

Conclusion: Mastering The Journey on The Digital Highway

The rapid evolution of digital technologies has fundamentally transformed the landscape of various industries and societal structures. The Digital Highway, as explored throughout this book, encapsulates the multifaceted nature of these transformations. This final section synthesizes the key insights and implications derived from the preceding chapters, offering a comprehensive understanding of this digital shift.

The analysis of digital infrastructure reveals that advancements in computing power, data storage, and network capabilities have collectively facilitated unprecedented levels of connectivity and data exchange. These technological underpinnings have enabled the proliferation of the Internet of Things (IoT), Artificial Intelligence (AI), and Blockchain Technologies, which have been instrumental in reshaping traditional business models and operational paradigms.

Economic implications of digitalization are profound. The integration of digital tools has led to increased efficiency, cost reduction, and the creation of new revenue streams. However, it is also accompanied by challenges such as job displacement due to automation, digital divide, and cybersecurity threats. Policymakers and business leaders must navigate these complexities to harness the benefits while mitigating adverse effects.

In the realm of social dynamics, digital platforms have redefined communication, social interaction, and information dissemination. Social media, in particular, has emerged as a powerful tool for both positive engagement and divisive misinformation. The dual-edged nature of these platforms necessitates a balanced approach to regulation and user education to foster a healthy digital ecosystem.

Educational frameworks have been significantly influenced by digital tools, facilitating remote learning and personalized education. The COVID-19 pandemic accelerated the adoption of digital education, highlighting both its potential and the existing disparities in access and digital literacy. Future educational policies must address these gaps to ensure equitable learning opportunities.

Healthcare systems are undergoing a digital transformation through telemedicine, electronic health records, and AI-driven diagnostics. These innovations promise improved patient outcomes and operational efficiencies. Nevertheless, the integration of digital health tools raises concerns regarding data privacy, ethical considerations, and the need for robust regulatory frameworks.

Environmental sustainability is another critical area impacted by digital technologies. Smart grids, precision agriculture, and digital monitoring systems contribute to more sustainable resource management. The environmental footprint of digital infrastructure, such as data centers and electronic waste, presents an ongoing challenge that requires innovative solutions and policies promoting sustainable practices.

The geopolitical landscape is also being reshaped by digital advancements. Cybersecurity threats, digital espionage, and the strategic importance of digital infrastructure highlight the need for international cooperation and robust cybersecurity measures. Nations must balance technological innovation with security considerations to maintain stability and foster global collaboration.

Ethical considerations in the digital age are paramount. Issues such as data privacy, algorithmic bias, and the ethical use of AI necessitate a framework that balances innovation with societal values. Stakeholders, including technologists, policymakers, and ethicists, must engage in continuous dialogue to address these ethical dilemmas.

The Digital Highway thus presents a complex interplay of opportunities and challenges. It underscores the necessity for a multidisciplinary approach in crafting policies and strategies that leverage digital technologies for societal benefit. The ongoing digital revolution requires adaptive governance, continuous innovation, and a commitment to ethical standards to navigate the future effectively. The insights provided herein aim to inform and guide stakeholders, as they traverse, the evolving digital landscape.

This journey will require not only technological advancement but also a deep commitment to ethical practices, societal well-being, and the long-term sustainability of our planet. The future is digital, and those who master the journey on "The Digital Highway" will be the architects of a new, interconnected world.

www.ingramcontent.com/pod-product-compliance
Lightning Source LLC
LaVergne TN
LVHW052257210726
843527LV00040B/550